Rogue Asteroids and Doomsday Comets

Rogue Asteroids and Doomsday Comets

The Search for the Million Megaton Menace That Threatens Life on Earth

Duncan Steel

Foreword by Arthur C. Clarke

John Wiley & Sons, Inc.

New York • Chichester • Weinheim • Brisbane • Singapore • Toronto

This text is printed on acid-free paper.

Published by John Wiley & Sons, Inc.

This publication is designed to provide accurate and authoritative information in regard to the subject matter covered. It is sold with the understanding that the publisher is not engaged in rendering professional services If legal, accounting medical, psychological, or any other expert assistance is required, the services of a competent professional person should be sought.

Library of Congress Cataloging-in-Publication Data

Steel, Duncan

Rogue asteroids and doomsday comets : the search for the million megaton menace that threatens life on Earth / Duncan Steel.

p. cm.

Includes bibliographical references and index.

ISBN 0-471-19338-0 (alk. paper)

1. Asteroids—Popular works. 2. Comets—Popular works. 3. Impact—Popular works. 4. Extinction (biology)—Popular works. I. Title.

QB651.S74 1995

363.3'4—dc20 94-23409

Printed in the United States of America

10 9 8 7 6 5 4 3 2 1

Contents

Foreword

I began the "Sources and Acknowledgments" of my recent factually based novel, *The Hammer of God*, with the words "My involvement with the subject of asteroid impacts is now beginning to resemble a DNA molecule: The strands of fact and fiction are becoming inextricably entwined." *Rogue Asteroids and Doomsday Comets*, of course, adds to that involvement, providing another twist to the entanglement.

Back in 1973, I opened *Rendezvous with Rama* with an account of an impact upon the Earth in 2077 and its awful consequences. Duncan Steel has repeated those words, which still ring true, here in Chapter 11. Two decades ago, we knew comparatively little about the effects of asteroids and comets upon our environment—for example, the Tunguska event of 1908 was a mystery: How could it have left no crater if it were a fragment of a comet?—but research in the intervening two decades has shown that, to a large extent, I got it right. I suggested there that humankind would be spurred into starting a perennial surveillance program to ensure that, post-2077, such an event would never again occur. That program I entitled Spaceguard. In serving on NASA's International Near-Earth Object Detection Workshop (a mouthful of a name for any committee), Steel pointed out that in *Rendezvous with Rama* I had foreseen the type of search program that the committee had been charged by Congress with recommending, although I had based my fictional search on radar methods rather than optical scanning. It was flattering that *The Spaceguard Survey* was adopted as the title for that recommendation.

In between times, remarkable leaps have been made in our understanding of impacts and how they affect life on Earth. Probably the greatest single impetus helping to make this a respectable subject of study was the publication in 1980 of the startling theory that the mysterious sudden demise of the dinosaurs was due to a massive asteroid impact. That hypothesis, developed by Nobel laureate Luis Alvarez and his geologist son, Walter, was based upon their discovery of a huge anomaly in the amount of the rare metal iridium in the geological stratum from the time of the extinction—rare on Earth, but not so rare in meteorites. That development is described herein, along with the more recent ideas that indicate that there have been many mass extinctions in the past few hundred million years caused by such impact catastrophes. To add another twist to the DNA molecule, my only *non*-science fiction novel, *Glide Path* (1963), is dedicated to "Louie and his colleagues." In the closing years of the war, I was responsible for running the prototype ground control approach (GCA) unit, which he invented, though we did not actually meet until ten years later.

Because the subject was now respectable, and suitable for discussion in works of fact as well as fiction, magazine stories aplenty started to appear. In May 1992, I was flattered to receive a letter from *Time* magazine asking me to write a short story for their special "Beyond the Year 2000" edition, to "give readers a snapshot of life on Earth in the next millennium." This was only the second piece of fiction to appear in *Time*, being preceded by a story from Alexander Solzhenitzyn, published in 1969.

With the just-completed *Spaceguard Survey* report in my hands, I had the factual material available in order to provide a sound basis for my story, which duly appeared (see *Time* magazine, Fall 1992, volume 140, number 27). But that short story was not the end of it.

Duncan Steel and I were born in the same county, Somerset in western England, which explains our similar accents. In the middle of 1992, my brother Fred had organized a space festival in the town of my birth, Minehead, to celebrate my 75th year, and Steel came back from Australia to give a presentation about the real-life Spaceguard. That, among other things, convinced me that *The Hammer of God* should be a full novel. This was brought home by a message that Steel sent me soon after he had returned to Australia and I to Sri Lanka. The chronology I spell out in the factual part of *The Hammer of God*, but briefly Steel's message pointed out that the recently rediscovered Comet Swift-Tuttle might be coming back to strike the Earth in 2126. The excitement engendered in the world's media over this possibility—possible with the state of our knowledge then, although we now know that the comet will miss our planet by two weeks—was more than enough to convince me that readers would be interested in a full novel. I was excited as well, because in *Rendezvous with Rama* I had set the arrival of the hypothetical Kali in the year 2110, just sixteen years ahead of the real Comet Swift-Tuttle.

If there is interest in a fictional account of humankind being saved from a calamitous impact, then surely there must also be interest in the facts upon which my writings are based. In this volume, Steel sets out those facts as we understand them today. The statistics are worrying in some respects. Many might say, "But we know of no one that has died from an asteroid impact." That's true, but then who has ever died from a thermonuclear explosion? Surely no one would claim that *they* are not dangerous?

Finally, I cannot resist the temptation to gloat over two other statements that I have made in my publications. First, with regard to the Tunguska event, in part of this book Steel describes how his view of the hazard of asteroids and comets differs from the mainstream belief of scientists working in the area, and in particular how he sees relatively small objects in the Taurid meteoroid stream as posing the major risk. Back in 1980, in *Arthur C. Clarke's Mysterious World,* I pointed out that the Tunguska object was quite likely a member of that stream, a suggestion for which I am pleased to see I am duly credited in Chapter 9!

Second, as I write, the astronomical world waits with bated breath for the impacts of the fragments of Periodic Comet Shoemaker-Levy 9 on Jupiter. Some have suggested—in error—that those impacts might be enough to cause nuclear fusion to initiate in the core of that planet, turning it into a second sun. Precisely such an event, though with a somewhat different scenario, was the climax of *2010: Odyssey II*, splendidly rendered in the movie version. Though I do not for a moment believe that this will actually happen* (Jupiter would have to possess ten times its present mass to become a sun, and that of any comet is quite negligible), I am not averse to generating a little alarm. For the case of impacts upon the Earth, a little alarm is what is needed to get Spaceguard funded and under way, thus helping to guarantee that humankind will see not just 2001, but 3001 as well.

ARTHUR C. CLARKE

Colombo, Sri Lanka
June 1994

**Author's Note:* What actually transpired in these phenomenal Jovian impacts (see Epilogue) entirely confirmed Clarke's predictions.

Preface

My intention in writing this book has been to present the facts and our scientific beliefs (and uncertainties) about the terrestrial impact hazard due to comets and asteroids in an easily understood way. I strongly believe that this is such an important subject that the public needs to be properly informed so that educated decisions may be made about how we should be addressing this hazard. Clearly, I am of the opinion that this is a problem that can and should be not only tackled, but also solved, because the future of the human race and all other forms of life on Earth may very well depend on doing so. Inaction at this stage is simply an indulgence in a game of Russian roulette, whether or not we know that the gun is pointing at us, and whether or not we realize that the trigger is being pulled. For the first time since life began on Earth, a species has the ability to ensure that life continues without a repeat of the impact-induced mass extinctions that have occurred in the past and will almost certainly recur in the future unless we intervene.

I have not intended to give a complete bibliography or to cite every single original reference that has provided a building block, no matter how small, to the citadel of knowledge on this topic; to have done so would have added considerably to the length of this book and would have been of little or no interest to most readers. I have, however, given references to some of the most pertinent or unusual publications that the keen reader may want to access, and have included suitable review papers for those who may want to pursue the details. The majority of the works cited should be intelligible to the interested layperson. For the reader who wants to delve even deeper, there is practically no limit to the amount of time that might be spent on following up the latest information on asteroids and comets. Original research papers running to many thousands of pages are published on this topic every year in a wide variety of scientific journals.

Of necessity, a book such as this contains certain technical terms with which the reader might not be familiar. These have been described, so far as is possible, in the Glossary or are defined as they are introduced in the main text. Although the reader will certainly be familiar with, for example, the word *comets*, the discussion in the Glossary might be of use in developing an understanding of how comets fit into the general scenario. It is hoped that unnecessary jargon has been largely excluded.

A large number of people have assisted me at various stages of the preparation of this book, either directly, with answers to my requests for information, or indirectly in discussions that may at the time have been seemingly unrelated. I have tried to make a list that is as complete as possible, but it is virtually certain that I have left someone out—my apologies. Anyway, I have much appreciated, in no particular order, the help of Rob McNaught, Brian Marsden, Graff Williams, Victor Clube, David Asher, Mark Bailey, Gerhard Hahn, Graeme Waddington, Bill Napier, Annette Callow, Russell Cannon, Robyn Shobbrook, Arthur C. Clarke, Stephen Bain, Paul Davies, Ken Russell, Tony Beresford, Gene and Carolyn Shoemaker, David Levy, Tom Gehrels, Bob Dean, Roy Antaw, Rhonda Martin, Malcolm Hartley, Gordon Garradd, Paul Cass, Lewis Jones, Graham Elford, Colin Keay, Herb Zook, Glo Helin, Jim Scotti, David Rabinowitz, Gennadij Andreev, Richard Grieve, Janice Smith, John Campbell, Alan Gilmore, Pam Kilmartin, Andrew Taylor, Jack Baggaley, David Morrison, Clark Chapman, Steve Ostro, Alan Harris, Don Yeomans, David Nicholls, Vic Gostin, Jim Klimchuk, Paul Thomas, Chris Chyba, Kevin Zahnle, Chris McKay, Sidney van den Berg, Michael Rampino, Shin Yabushita, Ichiro Hasegawa, Andrea Carusi, Robert Jedicke, Ted Bowell, Giovanni Valsecchi, Howard Jones, Tony McDonnell, Rick Binzel, Paul Almond, John Hanson, Aletha Jones, and, of course, my wife Helen and son Harrison, for whom asteroid 5263 Arrius is named.

DUNCAN STEEL

Coonabarabran, Australia
December 1994

1

Fiddling While Rome Burns

It is difficult to overstate the almost unimaginable energy that is released when a massive asteroid or comet hits the Earth. Merely stating that the explosive power is far greater than all the world's nuclear arsenals combined does not properly convey matters. The reader may think that such combined power might simply result in a larger area being flattened than that which a nuclear bomb devastates. Instead of the holocaust wreaked in the few square kilometers of central Hiroshima, for example, we might imagine all the buildings in the metropolis of Los Angeles being toppled. In fact, the impact of a large asteroid or comet is quite different from that. Were one to land in Southern California, for example, all of Los Angeles along with several kilometers of the rock from the Earth's crust beneath it would be picked up and largely vaporized, lumps raining down on Hawaii and New York an hour or so later. Not that Honolulu or New York City would be left standing by then. Phenomenal seismic shocks following the impact would have already shaken them flat.

This is a terrifying scenario, certainly. But how likely is it? Scars left on the Earth's surface suggest it is far from science fiction. There are, in fact, many enormous craters on the surface of the Earth that have been conclusively identified as the fossils of just such impacts. The best-known crater in the world is Meteor Crater in northern Arizona. Twelve hundred meters wide, 170 meters deep, and 5 kilometers in circumference, Meteor Crater is the surviving scar from an iron meteorite thumping into the plains thereabouts around 50,000 years ago in a cataclysmic explosion that liberated energy equivalent to 20 million tons (Mt) of TNT. To put that number into perspective, consider that the explosion that formed Meteor Crater was equivalent to almost 2,000 times the power of the bombs dropped on Hiroshima and Nagasaki.

Impressive as it is, Meteor Crater is a relative newcomer, and a pipsqueak as craters go. Just a few years ago, another large impact scar—180 kilometers across—was identified on the Yucatán Peninsula in Mexico. The impact that left this crater is thought to have been capable of such damage that scientists theorize it provides the answer to the long-standing mystery concerning the cause of the dinosaur extinction 65 million years ago. Based on evidence in the fossil record, it seems that the environmental disaster wreaked by this impact was of such global proportions that it caused the extinction of a large fraction of the species then extant, including the dinosaurs, and heralded the beginning of the Tertiary era, and, of course, the ascendancy of the mammals. We will learn more about this crater, and this theory, later in the book. We will also see that the geological record shows plentiful evidence of other global environmental upheavals that can be linked to large impacts by asteroids and comets. It is only in recent years, however, that we have known how to identify much of this evidence.

The irony is that after spending billions of dollars sending men to the Moon and studying the craters there, we have at last realized that there are plenty of craters to study that are accessible at a rather more modest cost. There is, however, good reason that more craters haven't been identified sooner. For decades, maps of South Australia have shown a 35-kilometer-wide dry salt pan called Lake Acraman, but it was only in 1986 that geologists realized that this is the residual basin surviving from a colossal impact about 600 million years ago, forming a crater 90 kilometers across. The true nature of the lake was unrecognized for so long because the telltale signs of an impact have been worn away. Over the past 600 million years, the effects of wind, rain, and ice have gradually chipped away at the shape of the original crater. Other remnants of the impact were, however, left in the geological record some distance away. The collision that created the crater also caused a shower of rock to be disgorged from the site, which then rained down more than 300 kilometers away in a region where the mountains known as the Flinders Ranges now stand. At the time, that region was occupied by a shallow sea, so that the ejected rock accumulated in the sediment laid down there over eons, eventually buckling up to form the mountain range. We have been able to identify Lake Acraman as a crater because of the ejected rock fragments found in the Flinders, which are so different from the predominant local geology but are identical with the rocks at Lake Acraman.

Indeed, due to the obscurity of many of the craters on Earth, much of our initial knowledge about how craters are formed came not from the study of terrestrial craters, but from evidence gathered from studies of the lunar surface. If the Earth could be said to be pockmarked by craters, then the Moon appears to have had a more severe dose of the same ailment. Even a brief peek through a very modest telescope, or a perusal of satellite images, indicates that our lunar companion has suffered numerous and frequent impacts over the eons. In fact, the Moon's surface is so saturated with craters in many areas that any new incoming asteroid or comet will almost surely obliterate an old crater in order to produce a new one.

The origin of the lunar craters was a matter of dispute among astronomers through the ages, and their cause has only recently been settled, lending crucial support to the theories of the effects of impacts on Earth. Even at the start of the Apollo program just three decades ago, it was still the favored idea among many geologists and astronomers that the origin of the lunar craters was volcanic action, despite the fact that evidence from telescopic studies should have made it clear that impacts must have been responsible. For example, many craters show a central uplift or spike reminiscent of the surface profile in a cup of coffee after a sugar cube has been dropped into it—a rebound occurs, the coffee spurting upward. In the case of the huge energies and pressures generated in a big impact, the rock target behaves for a while as a fluid, but is then frozen into this characteristic rebound form—a sure sign of an impact crater.

The case was sealed, however, when the Apollo rock samples were inspected. First, metamorphism in the rocks, which is indicative of the phenomenal pressures generated in impacts, was identified immediately. In addition, study of the rocks showed that they were pitted and cratered throughout by telltale signs of energetic impacts; that is, dust particles too small to be seen with the eye had zapped into the Moon, causing tiny craters, around and within which the rock was melted to leave a glassy surface. Slightly larger projectiles, perhaps the size of a pinhead or more, had produced craters with dimensions of thimbles; such projectiles are called *meteoroids*. Indeed the whole of the lunar surface had clearly been turned over many times by relatively small impacts by such meteoroids, from pinhead to basketball size, a process that selenologists call *gardening*. And this process is still going on today on the Moon.

This discovery led to the obvious question: So why, apparently, didn't (and doesn't) a similar steady bombardment happen on Earth? Why was the Earth not similarly saturated with craters and why are we not constantly assailed by meteoroids? A quite simple calculation shows that the number of impacts per unit area of the Earth should, in fact, be about twice that of the Moon, because the Earth's larger gravity should pull more impactors in as they orbit the Sun. According to that calculation, the Earth should have 25 to 30 times the number of craters in total. Where are they? As in the case of the Lake Acraman crater, we now know that most of the craters formed on Earth must have been obscured. And as for the steady bombardment of meteoroids, we are shielded from them by the Earth's atmosphere.

The reasons that Earth craters are not obvious, and that we are not daily peppered with meteoroids like buckshot, are explained by the very different environment here on the Earth, compared with the Moon. We are well aware that the terrestrial water and oxygen supplies have made it possible for life to proliferate here and for you to be able to sit and read this book. But the air is also responsible for protecting us from the small meteoroids that would otherwise rain down, and the water is accountable for obliterating much of the evidence of large impacts that would otherwise have made clear to us our finite mortality—as a species, if not as individuals.

Meteoroids hit the lunar surface unimpeded by any atmosphere; shooting stars cannot be seen from the Moon. And once meteoroids strike the Moon, the craters that are produced remain as scars for eons, obliterated only if another chance impact happens in the same location. The Moon also has no rain or wind to erode away the craters. It has no continental drift and no active volcanoes, which have also played a part in obliterating or obscuring the craters on Earth.

The Moon has an environment friendly to incoming meteoroids, asteroids, and comets in that they reach the surface intact. The converse is true of Earth. A meteoroid approaching the Earth, maybe the size of a nut and bolt, starts to heat up as it meets the upper atmosphere. At an altitude of 120 kilometers, where the atmospheric density is only one ten-millionth of that at the surface, friction causes this heating, pounding the material into gas particles. By 100 kilometers' altitude, the material has heated up so intensely that it is not only melting, but also boiling off, producing a glowing trail, which you may witness as a shooting star. By 80 kilometers, there is practically nothing left of the meteoroid, and the shooting star has had its short burst and dies out, usually after just two or three seconds.

The larger the meteoroid, of course, the deeper it may penetrate into the atmosphere. Given a suitable composition (nickel-iron), density (high, around eight times that of water), a shallow-entry angle (so that it slows gradually, having a longer path through the upper atmospheric layers), and a low entry speed (11 kilometers per second being the lowest possible), a meteoroid the size of a basketball may reach the surface, although only a fist-sized chunk might remain. This could perhaps cause some damage or even injury—there have been a few cases in recorded history of humans being hit, and property damage occurs about once a year on average.

One particularly impressive meteoroid entry happened on April 9, 1993, when a very bright meteor—what is known as a fireball—was seen by hundreds of people over the south coast of New South Wales in Australia. Most had never seen anything like it before, and it engendered great excitement, to say the least. At the time, I told the media, who were clamoring for information, that such an event might be seen from any one spot only once every few years, maybe once every decade—and then only if you stayed up all night, every night, and kept your eyes open. It was therefore a source of some embarrassment when precisely a week later, an even bigger meteoroid—estimated at about 3 to 4 meters in size—arrived in the atmosphere high above the Queensland border, zipping across the sky on its southerly path across

New South Wales as it turned night into day for a few seconds, eventually petering out at a height of around 18 kilometers above the small city of Dubbo. On its supersonic path, it generated a shockwave felt in Dubbo and throughout the surrounding area to a radius of at least 100 kilometers. By the time that I got through to the main Dubbo police station switchboard 30 minutes later (to tell them what had happened), the operator had logged over a hundred calls from people claiming variously that a bomb had gone off, that a jet had flown just over their rooftops, or that someone or something was smashing into their house—a hundred calls at that station alone, with thousands more received elsewhere. It seems that many of the rest of Dubbo's 30,000 inhabitants were hiding under their beds, and the State Emergency Services had been alerted as houses were shaken to their foundations and windows vibrated. The energy released in the detonation was roughly equivalent to that released by the Hiroshima bomb, so the thought that a bomb had gone off was not so bizarre. This "bomb" went off, however, 18 kilometers above the city, which is close to twice the height at which a jumbo jet flies. And no meteorite reached the ground.

How could a mere lump of rock or ice produce such a gargantuan explosion? The answer is that at a speed of 30 kilometers per second, any solid body has a hundred times more energy available for destructive purposes than if it were made of pure dynamite. That sort of speed is outside the realm of our everyday experience: It is about a thousand times the highway speed limit in most countries and somewhat more than a hundred times the velocity of a jumbo jet. However, it is the speed with which we circle the Sun on planet Earth, imperceptably varying by a few percent between winter and summer. It also happens to be a reasonably good guess at the likely impact speed of an asteroid or comet smashing into our planet.

It is extremely fortunate, therefore, that the vast majority of potential impactors entering the Earth's atmosphere do not reach the ground. Even an asteroid as big as a city block will more than likely detonate high above the surface, and it is not until sizes larger than about 100 meters are reached that there is a strong chance of a cratering event. For 50-meter asteroids, the chance of reaching the surface intact is of the order of 1%.

So how many asteroids are there out there of truly dangerous size? If you walk along a beach and randomly pick up pebbles, you'll find that by far the largest fraction are small in size, with an ever-decreasing number of larger stones: For every one the size your fist, there are ten or twenty the size of your thumb, and maybe fifty the size of your little finger. The same sort of rule applies in space: For every grapefruit-sized meteoroid, there are a dozen apple-sized and a hundred pea-sized meteoroids. But, as we have seen, even a pea-sized meteoroid packs a punch. In fact, about the same punch as a stick of dynamite, using the "hundred times as much energy" rule. On a clear night you may see ten shooting stars (or meteors) per hour, their death throes illuminating the sky as they burn up 100 kilometers above your head. Those are due to meteoroids just the size of a pea. Once an hour you may see an especially bright one, perhaps due to a walnut-sized particle, and once every few hours a dazzling streak as a grapefruit-sized meteoroid meets its end in the tenuous air far above. A basketball-sized lump, giving a real humdinger, which may instantaneously light up the country around you, you'll see maybe once a month if you have the time and the patience. The question is, how long is it between the *big* lumps of rock or ice, which will punch through the atmosphere to wreak havoc below? The answer to that, according to our presently incomplete understanding, is about a century between 50-meter rocks potentially big enough to wipe out a city or even a good-sized European country or a small U.S. state (for example, New Jersey or Connecticut). Correspondingly, it is about 100,000 years between 1-kilometer asteroids and comets, which could kill a large fraction of humankind and send the rest back into a twenty-first-century version of the Dark Ages.

These chances may seem inconsequently slim, but given the evidence we now have that these cataclysmic events have happened in the past, and given a recent proliferation of findings of asteroids and comets that are on course to cross the Earth's path, we must consider very seriously the probability of future impacts. And our estimations need not be in the realm of the purely speculative. Indeed, in 1992, a great news frenzy was set off when a comet, which had been beyond the realm of our telescopes for 130 years, was rediscovered. Calculations of its orbit made at the time showed that it might well collide with the Earth on its next trip around. Nineteenth-century observations of that comet, discovered by American astronomers Lewis Swift and Horace Tuttle in 1862, suggested that it would be back in 1980–82, but it did not appear. Astronomers have known for more than a century that the comet—Comet Swift-Tuttle—has an orbit that intersects that of the Earth, because each August a meteor shower is seen, those meteors having been released from the comet on some long-forgotten transit through the inner solar system. But if the tiny meteoroids can hit the Earth, providing a spectacular fireworks display as they zip into the atmosphere at 60 kilometers per second, then surely the comet itself could do so. Calculations at the time of the rediscovery made by astronomer Brian Marsden, who runs the Minor Planet Center at the Smithsonian Astrophysical Observatory, showed that from the observations immediately available, Comet Swift-Tuttle would return in July or August of the year 2126. The Earth will cross the path that the comet will follow on August 14, 2126, so that a collision, although unlikely, was certainly possible, according to our knowledge at that stage.

How much energy would be released in such a collision? It is difficult to say. We know the speed at which the impact would occur, but we are unsure of the mass of the comet, which is a crucial variable in calculating the energy released. A reasonable estimate, however, is an explosion equivalent to about 100 million megatons of TNT.[1] It's hard to grapple with a figure of that enormity. What does it mean? *Mega* does not explicitly mean "big," as in the term *mega-hit*, often used in promoting pop music. In scientific parlance, *mega* means "a million," so here we are talking about taking a hundred million million tons of TNT, and setting it off all in one place and all at one time—a far bigger explosion than humankind could muster using the united nuclear arsenals of all nations. In terms of the effects of a nuclear weapon with which we're all familiar, it's equivalent to setting off a bomb a billion times as powerful as that dropped on Hiroshima. Alternatively, think what would happen if we detonated a Hiroshima-scale bomb over every 2 square kilometers of the Earth's surface, populated and desolate, land and ocean, arctic, antarctic, and tropical alike. What might survive? Certainly our civilization would not, and maybe not our species either.

A crude estimate of the chance that Comet Swift-Tuttle would indeed collide with the Earth in the twenty-second century, ending life as we know it (although cockroaches and some deep-sea creatures would survive), puts the probability at about one in 10,000—good odds for anyone wanting to play Russian roulette, but newsworthy? Look at it this way: If there are 10 billion people alive in 2126 (likely a gross underestimate), then if the chance of their all dying were one in 10,000, it follows that the expectancy of deaths calculated at that stage—that is, with the state of our knowledge in 1992—is a million people being killed. In fact, it would be all or nothing, but a million is the *expectation value*, calculated in the same way as an insurance company might determine that you have a one in a hundred chance each year of wrecking your $20,000 car, rendering an expectation value of $200 worth of damage, resulting in their charging you a premium of $300 or more. A million people is equivalent to a jumbo jet crashing and killing all its passengers and crew every two or three weeks for the next 130 years—the time it will take for the comet to come back again. That's newsworthy. That's why there was a hoopla.

So will the comet collide with the Earth in 2126? As chance would have it, a handful of scientists, working individually and separated by thousands of kilometers, put together the evidence that eventually led to sighs of relief when it was realized that the comet will *not* hit our home next time around. Back in 1973, Marsden had published a paper pointing out that Comet Swift-Tuttle, according to his analysis of the data available from 1862, was due back in about 1980, so that astronomers should start looking for it in the late 1970s. However, Marsden also pointed out that, while in China in 1737 as a Jesuit missionary, Ignatius Kegler had seen a comet that might have been Swift-Tuttle on its previous apparition. If this were the case, then the comet would not reappear until 1992. When the comet was recovered in 1992, Marsden was proven correct, and that led to the recognition that it would come perilously close to the Earth in 2126. What was needed were more observations of the comet, from 1993 onward as it moved away from the Earth, getting progressively dimmer while passing deep into the southern sky, beyond the reach of all northern hemisphere observatories.

After Marsden pointed out that an impact was possible, based on the 1737 and 1862 data, another expert in celestial dynamics, Don Yeomans, of the Jet Propulsion Laboratory in California, also looked at the problem. Yeomans and Marsden had worked together many times, and had been friends for some years, but soon an unholy argument broke out. Yeomans decided that one set of observations from 1862, made at the Cape Observatory in South Africa, were slightly in error and so should be rejected. Using the other measurements of the comet's position night by night from 1862, Yeomans found that his computer program predicted a return date in 2126, which would lead to the Earth's being comfortably missed. Conversely, Marsden held that the South African data were accurate and indicated that the comet was subject to a sporadic nongravitational force (due to the jetting effect of water and other volatile chemicals evaporating from the cometary nucleus).

More data were required to resolve the argument. However, even modern, accurate measurements of the positions of Comet Swift-Tuttle as it recedes from the Sun were known to be unlikely to provide an answer. What would be useful would be ancient records of the comet passing through the inner solar system. Going backward from 1737, the previous apparition would have been so far from the Earth that a telescope would be necessary to observe it; that apparition occurred in 1610, around the time that Lipershey (in Holland) and Galileo (in Italy) were inventing such devices, so that it could not have been observed. By chance, *none* of the previous ten or so apparitions could have been seen, because the comet would have been too distant from the Earth to have been picked up without a telescope. But it was possible that it could have been detected in A.D. 188; at that time it would have been close enough to have been a naked-eye object.

For more than 40 years, Japanese astronomer Ichiro Hasegawa has made a serious pastime of studying ancient Japanese and Chinese accounts of celestial events, looking in particular for records of comets or meteor showers. Sure enough, back in the early 1970s, he noted that in records from A.D. 188 there was mention of a comet that would fit in with the location in the sky where Comet Swift-Tuttle might have been expected to appear. The previous apparition was unfavorable, so no record was to be expected. But the one *before* that was a possibility, and in the calendars of celestial phenomena observed in 69 B.C., Hasegawa found more mentions of what was perhaps Comet Swift-Tuttle. In fact, Marsden had suggested this to him a few year earlier.

Although these observations from two millennia ago are not precise in terms of exact positions in the sky, such as are those we now determine, knowing the dates and the approximate locations made it possible to try to calculate a solution to the motion of Comet Swift-Tuttle. Trying to fit a theory to the observations from A.D. 1992, 1862, and 1737, and 188 and 69 B.C. is no easy task, however, because the comet must be stepped backward in a computer day by day, taking into account the distances from each of the planets and hence the gravitational tugs from each. At Oxford University, Graeme Waddington found that it was possible to explain the comet's history (and therefore to extrapolate its behavior into the future) on the basis of a model constructed in this way. Others elsewhere, such as Gary Kronk in Illinois, arrived at the same result independently, and Marsden's assistant Gareth Williams also lent a hand to this little consortium. In California, at the Jet Propulsion Laboratory, Yeomans, Kevin Yau, and Paul Weissman also used Hasegawa's identifications to come to a very similar conclusion. The result: We can say when Comet Swift-Tuttle will be coming back in the twenty-second century to within two days, and it will miss the Earth by about two weeks. Indeed, it appears that this particular comet will not hit the Earth at any time in the third millennium; the first possible date is in A.D. 3044, but again, that's a long shot.

On this point Yeomans and Marsden are in agreement—that Comet Swift-Tuttle will not impact the Earth in 2126. They still differ, however, in their interpretation of the South African observations of 1862. Yeomans believes that they are in error, whereas Marsden points out that they are internally consistent and therefore point to some real behavior of the comet which is yet to be explained and understood. If nothing else, this indicates that there are always going to be things about which we cannot be certain (if such fine intellects as Yeomans and Marsden differ so strongly). It must be clear that they are not fools arguing over some trivial point; they are world experts in this particular field, and they are not arguing out of animosity (as, unfortunately, often occurs in science). This is a straightforward disagreement about how to interpret the facts, and it may well be insoluble. Similar instances may occur in the future as possible Earth impactors are identified, so there will always be some uncertainty and some place for opinion to be voiced. Science is rarely cut and dried.

If there is uncertainty, how can I state here with such confidence that Comet Swift-Tuttle will indeed miss the Earth in 2126? The answer is that a formal calculation of the possible errors in the observed positions for the comet indicates that the chance of its hitting our planet is less than one in a billion. Is that anything to worry about? Well, the threshold for the size of an impactor that could cause global effects is about 1 kilometer. An asteroid or comet of such dimension hits the Earth about once every 100,000 years, according to best estimates.

Therefore, in the 130 years before Comet Swift-Tuttle is due back, there is a better (or worse?) than one in a thousand chance that civilization will have met its Nemesis in the form of an alternative cosmic impact by an Earth-crossing asteroid or comet. I am not, therefore, so much worried about Swift-Tuttle as I am about some other possible impactor. If you are playing Russian roulette and you know there is a bullet in one of the six chambers, you tend not to worry unduly about whether the mosquito about to suck your blood might be carrying malaria. We are virtually certain that Comet Swift-Tuttle will miss the Earth in 2126; what we need to turn our attention to are the many Earth-crossing asteroids we have yet to spot, and which are more likely to give us a bloody nose soon. As we will see, there are, in fact, more and more potential impactors being identified all the time, and the risks posed by some of these are very much worth our attention.

The view expressed in this book is that searching out all large Earth-crossing asteroids is an important task that we *should* tackle, and that we *can* tackle. We have, over the past three decades, developed the technologies necessary not only to detect possible terrestrial impactors, but also to intercept and divert them. This technological development has arisen in parallel with scientific advances that have led to the recognition that we—meaning all varieties of life—are gravely threatened by possible impacts. The discoveries that have led to this recognition have been in many disparate fields: geophysicists and geochemists finding trace element and amino acid anomalies at geologic horizons; geologists and geomorphologists studying craters on the Earth, the Moon, and elsewhere in the solar system; astronomers finding asteroids and comets, which provide the ammunition for the crater-forming impacts; paleontologists investigating mass extinctions of species and their possible causes; and evolutionary biologists considering how changes in animal forms—and thus the proliferation of certain species and the demise of others—might be promoted by sudden, short-lived, cataclysmic alterations of the environment. The capabilities in astronautics, in computer systems and robotics, and in nuclear weaponry, which could be used to divert any object found to be heading our way, were certainly not developed to tackle rogue asteroids, but they may yet save us.

Technologically, we now have the capacity to ensure our long-term survival on the Earth, and thus, I am sure, elsewhere in the cosmos in the longer term. But are we willing and able to make the necessary economic commitment? There is no doubt that space missions to divert any identified menace would be very expensive, costing many billions of dollars and most likely requiring immediate and close cooperation between all space-faring nations. Given the circumstance that a prodigious impact is inevitable without our intervention, however, perhaps we should consider that expense is no object; in other words, "A horse, a horse, my kingdom for a horse."

Any interventionist action, though, awaits the positive identification of a definite, or at least highly likely, impactor for a simple reason: There is only a small chance that a large projectile is going to hit the Earth within the next few centuries, and even if there is a Nemesis rock out there, a suitable search program will almost certainly find it well ahead of time, with ample leeway (meaning several years at least) remaining for the appropriate response to be prepared. The first thing, then, is to perform the requisite search for the fateful Nemesis. If we do so, there will very likely be time after its discovery to build and equip a nuclear-armed interceptor fleet and send it to provide the small nudge needed to ensure that the offending object gives our planet a wide berth. The major expense of the development of such a fleet is therefore not a consideration at this stage, that "horse" certainly being purchasable when and if the time comes. We would have no choice.

It is easy to demonstrate here that it makes no sense *not* to carry out a suitable search program. A 1991–92 NASA study, the Spaceguard report, has shown that all potential Earth impactors down to 1 kilometer in size could be discovered and tracked in a program costing $300 million spread over 25 years. This size of impactor, and any larger, also happens to be the size that causes global destruction, which in our time would mean, at a minimum, the death of a large fraction of humankind, with those who survive being thrust back into a Dark Age existence. That $300 million is a lot of money. It could be used to provide housing for the poor or better health care, but it could also be viewed as a sort of insurance premium against people's losing their houses and their health (as the result of an impact). The question is: "How big is the liability?" This question can be answered easily, or at least a ballpark figure can be given for the appropriate figure. As will be discussed later in this book, a 1-kilometer asteroid impact could cause the death of at least 25% of humankind, and such catastrophes may be expected about once every 100,000 years. If we were to consider the United States alone, and value each American life at 3 million dollars (the lifetime contribution of each to the Gross National Product), then the annual expectancy of loss is about $2 billion. This is more than six times the cost of the proposed Spaceguard search program, that cost actually being spread over two decades. Not to carry out the survey is economic stupidity.

In the chapters that follow, I will first describe the threat that impactors pose, then the evidence that such devastation has been caused by impacts on the Earth in the past, and finally, in the last part of the book, the plans proposed for searching out and destroying incoming objects, if we decide to do so. The intention of this book is to provide the basic facts with which the concerned reader can determine whether we are, by not more aggressively addressing the impact threat, perhaps fiddling while Rome burns.

2

The Swarm around Us

The French philosopher and physicist Simon Laplace has often been credited with suggesting in the late eighteenth century that comets could have catastrophic effects on the Earth, either through an impact or a close passage by our planet.[1] In fact, it was Sir Edmond Halley who first offered this theory. Halley (1656–1742), a professor at Oxford University, made the novel suggestion that comets observed in 1456, 1531, 1607, and (during his lifetime) 1682 were one and the same; the fact that some comets come back repeatedly had not been recognized earlier. He predicted that the comet would reappear in 1759, which it did 17 years after his death. This was one of the triumphant predictions of Newtonian gravitational theory, Sir Isaac Newton (1642–1727) being an associate of Halley. As a result, the comet has henceforth been known as Comet Halley, although he did not discover it; *modern* naming rules apply the discoverer's name to any comet found. In the case of Comet Halley, there is a good reason for ignoring that rule: We do not know who discovered it, its first observation dating from the third century B.C.

It is to be expected that Halley would have perceived the possibility that the comet for which he had so accurately predicted the orbit might at some time in the future hit the Earth. Comet Halley's orbit crosses that of the Earth, which would make this prediction obvious. Halley, in fact, preceded Laplace by more than a century in making the suggestion about cometary impacts, but Halley's idea was suppressed by the Church. On December 12, 1694, Halley presented a lecture to the Royal Society of London (the main forum for new scientific results in Britain at the time), entitled "Some Considerations about the Universal Deluge." In this talk, he discussed how the story of the biblical flood may be an account of a cometary impact, with the ". . . vast depression of the Caspian Sea, and other great lakes of the world . . ." being formed in such cataclysmic impacts. This was no doubt a captivating idea, as it continues to be. Barely a month goes by without my receiving a letter from someone suggesting the concept anew; indeed, these correspondents may well be right. Many cultures have very similar flood myths, often involving some object arriving from the sky with catastrophic consequences. Halley's suggestion was too much for the Church in his day, however, because his explanation of the flood threatened the concept of divine intervention, and he was quickly pulled back into line. By the following week, he was giving another talk, entitled "Some farther Thoughts on the same Subject," in which he recanted his idea of a link between the biblical flood and a cometary collision, clearly under ecclesiastical pressure. In fact, another 30 years elapsed before Halley's presentations were published. Two years after Halley's lecture on cometary impacts, however, another Englishman, William Whiston, published a book in which he suggested that cometary impacts were important events in the Earth's history, although he did so in muted tones.

Early ideas about the effects of comets passing close by, if not actually impacting, the Earth were far off the mark, in part because astronomers had no sensible idea of the masses of comets. Indeed, we now know that the values estimated in the early calculations were too high by a large factor. For example, Laplace's idea of the devastation that would be caused was that a comet passing close by the Earth would raise huge tides on our planet through its gravitational pull on the oceans. In fact, in more recent times, the lack of any observable effect of this nature being produced by comets allowed a maximum value of the typical cometary mass to be derived, and this was much smaller than previously assumed. This more accurate calculation put to rest other erroneous ideas, such as that of French philosopher the Comte de Buffon, who in 1745 had proposed that the planetary system was formed after a comet hit the Sun, with a string of globules of solar material being ejected in a fashion similar to the rebound of a sugar cube after being dropped into a cup of coffee. Buffon imagined that the planets would then coagulate from these globules. However, we now appreciate that a comet plummeting into the Sun would have not even as great an effect as a single grain of sugar dropped into a cup of coffee, because the masses of comets are insignificant on this scale.[2] It should be noted, however, that despite more than two millennia of observations and the several spacecraft sent to Comet Halley in 1986, establishing it as our best-studied comet, we still are unsure of its mass by a factor of 3.

So, we see that the idea of comets having the potential to wreak havoc on the Earth has been around for centuries, and the debate about whether such impacts have occurred in the past—and could happen again at any time—has resurfaced from time to time throughout the centuries. However, during centuries past, only a handful of comets were known to be on Earth-crossing orbits, and because a simple calculation showed that there is only about a one in 300 million chance[3] that any particular comet would hit the Earth on any passage across our orbit, everyone could feel comfortable with the belief that cometary impacts would be many millions of years apart.

These comfortable calculations took no account of possible impacts by asteroids, however. This was due to the fact that up until the twentieth century, no Earth-crossing asteroids were known to exist. Comets have always been easy to spot. This is because of the huge clouds of vapor—called *comae*—that surround them. Comae are produced by solar heating as they approach from the frigid reaches of space. If you take a cup of water and turn it into vapor, you form a steam cloud, and a similar process of evaporation happens to a comet as it moves closer to the Sun. A cometary coma is similar to the clouds with which we are all familiar, except that it may be even larger than the Earth. Being so large, these cometary clouds reflect a lot of light, which is why comets are so easy to spot. It is the light reflected by their comae that we really see. No wonder then that the ancients spoke of wondrous visions in the sky, moving from one place to another and causing general consternation—with no city lights to flood the sky, much more was seen than is witnessed by the average town or city dweller today. Nor is it surprising that Comet Halley has been seen every 76 years for more than two millennia, since at times it has appeared brighter than the planets and is quite notable for its tail and its comparatively swift motion when approaching the Earth.[4]

Although comets have been visible since time immemorial, the same is not true for asteroids. These have proved to be elusive. In fact, it was almost two centuries after the invention of the telescope before the first asteroid was discovered in 1801; this was the biggest asteroid that has ever been discovered—almost 1,000 kilometers across. As the nineteenth century progressed, dozens more asteroids were discovered, but all were safely traveling in a band of orbits between Mars and Jupiter. The German astronomer Johann Bode had invented an arithmetical progression which seemed to fit the distances from the Sun of all of the known planets, but with an unfilled gap between Mars and Jupiter. These asteroidal discoveries therefore appeared to fit into the concept of an orderly solar system and were explained as being fragments of an exploded planet. Because these asteroids were on highly predictable, near-circular orbits that would not cross the Earth's own orbit, no panic was prompted by their discovery.

This sense of security was, however, somewhat tested in 1932 when two new asteroids were discovered, which we now know as 1221 Amor and 1862 Apollo. Amor has an orbit that crosses that of Mars and comes close to the Earth, although it cannot hit us. This is known to be true in the present epoch, meaning the next few tens of thousands of years, although beyond that period Amor's orbit might change. The orbits of all planet-crossing objects change slowly under the influence of distant planetary perturbations—for most asteroids and comets, Jupiter has the dominant effect—but rapid or abrupt changes are possible whenever an asteroid or comet makes a close approach by a planet due to the strong gravitational tug imposed at short ranges. It is now common practice to play "cosmic billiards" with spacecraft based on this effect. NASA's *Galileo* probe, which reaches Jupiter in 1995, underwent deliberate accelerative encounters with Venus (once) and the Earth (twice). Asteroids and comets randomly undergo such close approaches, which may either accelerate or decelerate them, depending on which side of the planet they pass. Thus it is possible that Amor, or some other asteroid, may one day fly too close by Mars and be diverted into an orbit making an Earth impact possible. For the foreseeable future, however, the chances are infinitesimal that Amor will cause us any worry.

Apollo, on the other hand, was found to have an orbit that does cross the Earth's path, making an impact quite possible. If Apollo were alone, there being no similar asteroids, it could then be considered to be like the comets and have only a one in 300 million probability of hitting us during each orbit. The danger to the Earth would not be markedly increased, and there would be no real cause for worry. But over the next few years, another asteroid was found by chance, named 2101 Adonis, and then another, in 1937, named 1937 UB Hermes. There was a hiatus in discoveries imposed by World War Two, but during the latter part of the 1940s, several more Earth-crossing asteroids were found. One notable asteroid discovered at that time was 1566 Icarus, which has an orbit passing even closer to the Sun than does the planet Mercury.[5]

By the 1950s, asteroids on Earth-crossing orbits were being found with an alarming frequency, and each discovery made it even more likely that thousands more were still to be found. The asteroids discovered were typically a few kilometers in size and would release a million megatons of energy or more should they hit the Earth. Unlike comets such as Halley, these asteroids had orbital periods of only a few years, meaning that they had an opportunity to collide with the Earth much more often. Instead of an impact every 100 million years, a more appropriate figure was once every 100,000 years or so. And that was only the comparatively few large asteroids that could be spotted with the technology available some decades ago; it was to be anticipated that there would be myriad smaller objects that could smash into the Earth much more frequently, with less extensive but still catastrophic consequences. Given all of these discoveries, some astronomers began to make the first truly disturbing estimates of the actual magnitude of the impact hazard faced by humankind.

Two of the first scientists to give an account of how often the Earth might be impacted, in light of the most recent asteroidal discoveries, were the American astronomers Fletcher Watson and Ralph Baldwin. In 1941, Watson made an estimate of the impact rate upon the Earth based on only the first three Earth-crossing asteroids to be discovered. He realized that these three composed only the vanguard of many hundreds of such bodies yet to be discovered, meaning that terrestrial impacts must occur at least on a million-year timescale, and the lunar craters were similarly explicable. One might have thought that, given the flurry of discoveries, his calculations would have been eagerly received.

In actuality, they were met with skepticism. This was largely due to the problem that if his calculations were on the mark, there should have been quite a number of terrestrial impact craters that would verify this impact activity in the past. The problem was, where were the impact craters that might be expected to remain from ancient impacts? As we learned in the first chapter, these craters are in most cases difficult to identify.

Although Meteor Crater in Arizona is now accepted as an impact crater by even the most skeptical, as late as 1945 the experts at the U.S. Geological Survey refused to acknowledge that it was indeed the scar left by an energetic impact. This was in part due to the fact that no large meteorite had been found there. As would become clear in

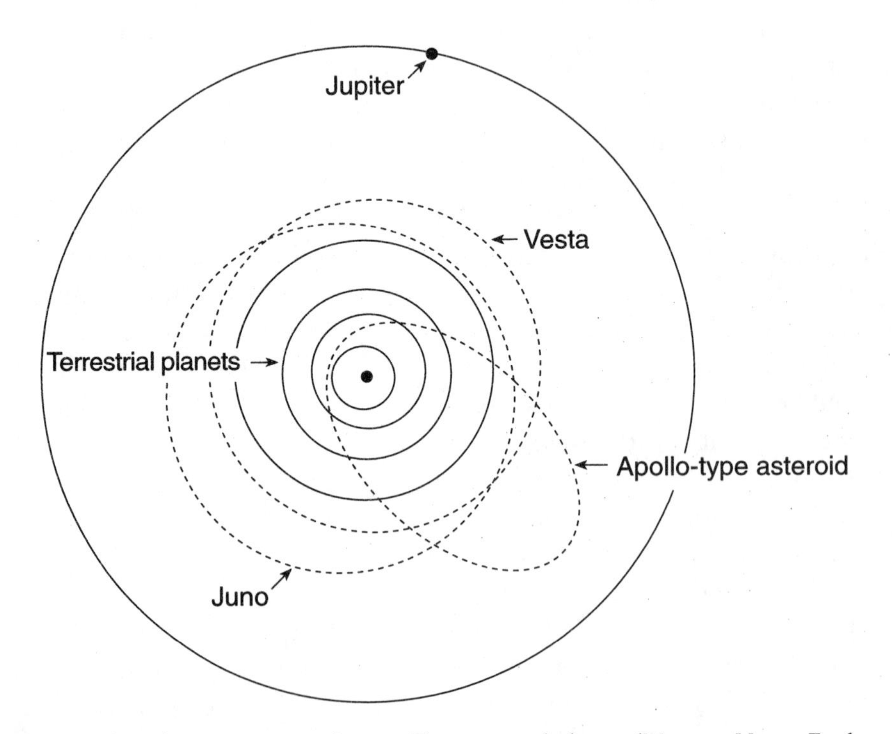

Figure 1. The solar system out to Jupiter. The terrestrial planets (Mercury, Venus, Earth, and Mars) are all relatively close to the Sun. In the large gap between Mars and Jupiter, there are millions of minor planets in a band called the *asteroid belt*. These do not cross the orbits of any of the planets, with the result that their own orbits are relatively stable. Two of the largest are shown here: Vesta and Juno. Also shown is a generic Apollo-type asteroid orbit, this one crossing the paths of Mars, Earth, and Venus.

future years, for reasons that will be explained later in the book, it is entirely to be expected that the impactor would not survive intact. At the time that Watson proposed his impact probability calculation, however, the origin of Meteor Crater, and many other craters, was still very much in dispute. Support for Watson's calculations mounted during the 1940s as more impact craters were recognized and their impact origin confirmed. In fact, well over one hundred impact craters (or impact structures in many cases, meaning the remains after extensive erosion) are now recognized on the Earth, with the number increasing annually. But discoveries about the *lunar* surface were also crucial to the eventual acceptance of impact predictions.

Over the years, many bizarre theories about the origin of the lunar craters had been suggested, such as the possibility that they were similar to the circular coral islands, or atolls, common in tropical oceans on the Earth. Considering that we now know that the Moon has always been waterless and lifeless, this idea seems stranger than it would have been when the nature of the lunar surface was unknown: The large dark areas on the near-side of the Moon are called *maria* (seas) because the ancients thought them to be oceanic expanses. Based on the resemblance between the lunar craters and some terrestrial volcanic craters, the favored explanation of the origin of the lunar craters was that they were the results of innumerable lunar volcanic eruptions. Even as late as 1979, I attended a conference in which there were strong arguments advanced for lunar craters being volcanic formations.

The notion that the craters of the Moon be taken as evidence of a history of impacting which would also be true of the history of the Earth was perhaps put forward most forcefully by Ralph Baldwin in his book, *The Face of the Moon*, published in 1949. Baldwin pointed out that because the Earth and the Moon were perennial companions on the same path around the Sun, the cratering history of the Moon suggested a disturbing notion: If the lunar craters were not all ancient, having been formed by a population of impactors which had since been dissipated, then similar impacts could recur. With the evidence of the newly discovered Earth-crossing asteroids with orbits such as that shown in Figure 1 (half a dozen were known at the time Baldwin was writing his book), as well as the comets like Halley that were known to be possible impactors, Baldwin was able to remark in regard to the most famous Moon impact scar, known as Tycho Crater, that: "The explosion that caused the crater Tycho would, anywhere on Earth, be a horrifying thing, almost inconceivable in its monstrosity." With this compelling warning, the modern era of terrestrial catastrophism began in earnest.

Continued discoveries of possible impactors have increased our knowledge to such an extent that we are now able to calculate with some reasonable degree of certainty the likelihood of an impact, by either a comet or an asteroid, in the foreseeable future. Indeed, we have learned that there is a veritable swarm of asteroids and comets with Earth-crossing orbits, as is depicted in Figure 2.

Not all of these objects need cause us alarm. We know that Comet Swift-Tuttle, for example, is not due to collide with the Earth within the next millennium. Due to the richness of data about the orbit of Comet Halley, we are also certain that it will not smash into the Earth within this time frame. There is a reasonable chance, though, that one of several other comets on Earth-crossing orbits, or a smaller object, most likely an asteroid, could wreak havoc in the foreseeable future. Before going further, we need to define what "the foreseeable future" means. For P/Halley and P/Swift-Tuttle,[6] we can be sure that we are

Figure 2. The congested route of the Earth. This plot shows the orbits of the 100 largest known Earth-crossers. The dot at the center represents the Sun, the dashed ellipse in the virtually clear central region showing Mercury's orbit. The outermost dashed line is the orbit of Mars. The orbits of Earth and Venus are plotted, but they are swamped by the lines representing the asteroids. This is a little misleading in that objects coming close to the Earth are more likely to be discovered; but then again, the plotted orbits compose only about 5% of the total of Earth-crossers. (Courtesy of Rick Binzel, Massachusetts Institute of Technology)

safe from an impact for at least the next thousand years because these comets come around only once per century or so, and they have been observed for more than two millennia, allowing us to determine their orbits with precision and thus to be able to predict their future evolution. It is therefore possible to say with confidence where they will be at least through the year A.D. 3000. For many other comets, however, and for the more recently discovered asteroids, the prediction of their future behavior is much more imprecise. Most asteroids have been observed for only a few decades at most, and they make frequent approaches to the planets. Because it is not possible to determine their orbits accurately enough to calculate the exact miss-distances in such encounters, it is impossible to predict their positions with confidence for more than 200 years hence. Thus "the foreseeable future" means the next couple of centuries. At the present stage of knowledge, it is believed that somewhere between 2% and 30% of the potential impactors that would cause a major impact are comets, with the rest being asteroids. The most likely type of catastrophe, then, is an Earth-crossing asteroid hitting the Earth.

What, then, given the present state of knowledge, are the chances for the occurrence of a massive impact? Might we expect one next year, or not for a million years? A proper calculation of collision probabilities between asteroids, comets, or meteoroids and any planet is quite complicated, with numerous factors to be considered, but a reasonable understanding of what is involved can be gained from the following thought experiment.

Imagine that you are in a room with a number of other people, all of whom are blindfolded. All of the people in the room are moving around in random directions at different speeds. You want to know the probability that one will collide with you. Obviously, the more people in the room, the higher the probability of a collision. Also, if the room is small, more collisions will occur, whereas if it is vast, collisions are much less frequent. These first and second considerations can be combined into one: The probability of collision varies with the number of people per unit area in the room. Third, if everyone stands still, the probability of collision equals zero, whereas if everyone is moving quickly, the probability of a collision becomes large; that is, the probability depends on the relative speeds of the other people compared with you. Fourth, if you are fat, you present a large target and a collision is more likely, whereas if you are thin, you are a smaller target and the probability of collision is smaller. All of these factors combined control the probability of a collision between you and one of the blindfolded people randomly moving around the room.

The calculation of the probabilities of the different kinds of near-Earth objects hitting our planet is formulated in a similar way by calculations we need not go into in detail here, but a table of relative probabilities is given in Table 1. The probabilities must be calculated separately for the different types of object. It is important at this point to define the different categories of comets and asteroids, and to describe in a general way the different characteristics of their orbits, because these differences contribute to the variations in the chances of each type of near-Earth object's actually hitting the Earth. There are two basic types of Earth-crossing asteroid. The first is the Atens, which

Table 1
Characteristic Collision Probabilities with the Earth for Different Classes of Object

	N	*P*	*T*	*P/T*	*V*
Asteroids					
Atens	12	16	0.85	19	15
Apollos	116	14	2.25	6.2	17
Comets					
Short-period	12	9	6.3	1.4	22
Intermediate-period	13	10	90	0.11	50
Parabolic	411	3.5	—	—	55

N is the number of known objects on which the estimate is based; I should point out that we know that the actual populations yet to be discovered are much higher. *P* is the mean collision probability with the Earth, per billion asteroid/comet orbits, for the sample *N*. *V* is the characteristic impact speed for that group, in kilometers per second. *T* is the characteristic orbital period (in years) for each group. So, if we take *P* and divide by *T* we get the impact probability per year. Note that no period is appropriate for the parabolic comets, which mostly pass through the planetary region once and are never seen again. Note also that *P/T* (the collision probability per year) for individual objects is highest for the Aten asteroids.

have an orbital period—meaning the time it takes them to make one complete orbit—of less than one year. The other type is the Apollos, which have orbital periods greater than one year. As for comets, there are three types. Short-period comets are those with orbital periods of less than 20 years. These are also termed Jupiter family comets. Intermediate-period comets, which are also known as Halley-type comets, have orbital periods between 20 and 200 years. And long-period comets have orbital periods in excess of 200 years, most significantly longer. In fact, only a few parabolic comets have ever been observed to have made a full orbit. Long-period comets are often called "parabolic" because they tend to be close to what is known as the parabolic limit, meaning the limit of orbital energy within which an object is gravitationally bound in an orbit around the Sun. Examples of these types of cometary orbit are shown in Figure 3.

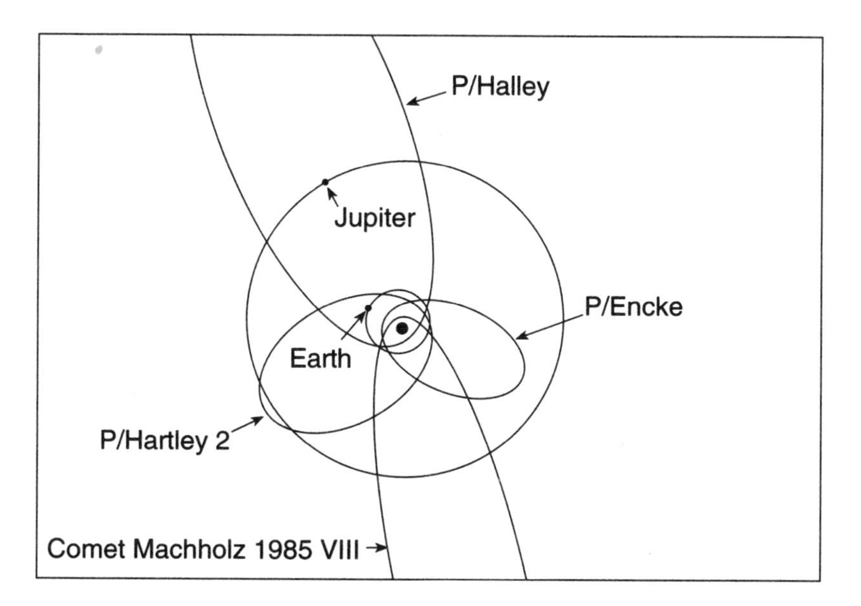

Figure 3. Many comets have Earth-crossing orbits. Shown here as examples are the orbits of a typical long-period, or parabolic, comet (Machholz, 1985 VIII), an intermediate-period comet (P/Halley), a Jupiter-family or short-period comet (P/Hartley 2), and the sub-Jovian comet P/Encke. P/Hartley 2, which was discovered by my co-worker Malcolm Hartley, is the proposed target of the European Space Agency's Rosetta mission.

Some additional terminology that will come in handy shortly are the terms used to describe the exact orbital characteristics of particular objects. Orbital distances are measured in what is known as the Astronomical Unit, abbreviated as AU, which is simply the average (or mean) orbital distance of the Earth from the Sun. This distance is approximately 150 million kilometers. And finally, the orbits of objects are often specified by what are known as the aphelion and the perihelion of the object's orbit. The aphelion is the furthest point in the object's orbital path from the Sun, and the perihelion is the closest point in the orbit to the Sun. Now, turning to Table 1, we can see the differences in the impact probabilities of these different types of objects.

When collision probabilities are calculated in this way, it is found that most are of the order of one per 100 million per orbit (of the object)—see Table 1. However the probability for some objects is much higher, and these tend to be those with perihelion or aphelion near 1 AU, in other words with one or the other orbital extreme close to the Earth. The main point is that not all asteroids and comets are equal, from the perspective of the hazard they pose. For example, compare 1862 Apollo and 2062 Aten, whose orbits are shown in Figure 4. Aten clearly poses a greater danger.

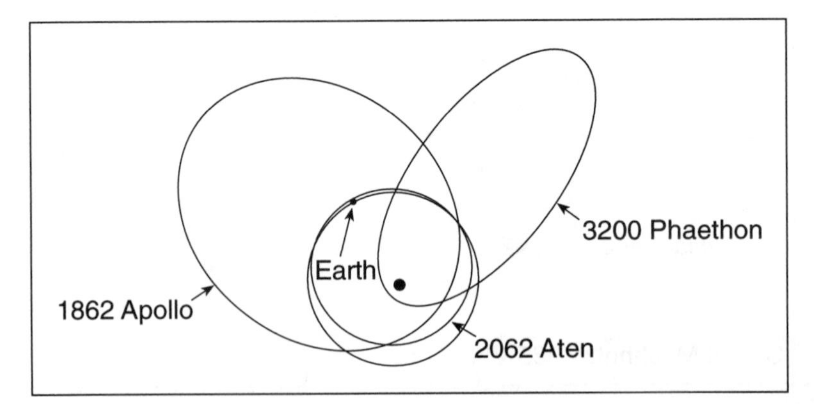

Figure 4. The orbits of 1862 Apollo, 2062 Aten, and 3200 Phaethon. Both Apollo and Aten (the archetypes of their classes) have orbits of moderate eccentricity, the former crossing the Earth's orbit as it approaches perihelion, the point in its orbit closest to the Sun, the latter as it approaches aphelion, the point in its orbit furthest from the Sun. Other Apollos and Atens, such as the generic orbit shown in Figure 1, have larger eccentricities. A specific example shown here is 3200 Phaethon, this body having the smallest perihelion distance of all known asteroids.

We have just summarized the probabilities that different types of near-Earth objects will hit the Earth, but in order to determine more precisely how frequently impacts on the Earth will actually occur, we next have to calculate the total populations in each class of object. Remember that our estimates so far have been based on only the number of objects of each type actually observed to date, and there are certainly significantly more objects yet to be found, of some types anyway. Before moving on to explain how we can better estimate the numbers of each type, it should be pointed out that these are *not* distinct classes with absolutely no overlap. Over the past few years, there has been mounting evidence that some, if not all, comets evolve into asteroids, and thus some Earth-crossing asteroids may originally have been comets. There are some astronomers who believe that the majority of such asteroids are extinct or dormant comets, myself among them. The reason for some of the confusion lies with the historical origins of the names of the classes, asteroids appearing starlike through a telescope and comets being diffuse. An extinct comet is one that has become totally de-volatilized, so that it will appear asteroidal into perpetuity; a dormant comet is one that has formed an insulating crust, preventing any substantial loss of volatiles, so no coma is apparent. In 1992, it was realized that an "asteroid" discovered in 1979 and known thereafter as 1979 VA had actually been observed earlier, in 1949, and was named Comet Wilson-Harrington because the discoverers detected a weak coma and tail. Apparently, in the intervening years this outgassing had ceased. This so-called asteroid is now known as 4015 Wilson-Harrington, but no one knows whether it will later burst into life once more. The outer solar system object 2060 Chiron showed no cometary activity when it was discovered in 1977 and so was numbered and named as an asteroid, but in the late 1980s, as it approached perihelion, it started to form a coma and tail—clearly a comet. Several asteroids have been found on orbits that are more typical of comets (for instance, having Jupiter-crossing paths, which are unstable). An extreme example is 5335 Damocles, which was discovered by my colleague Rob McNaught in 1991. This asteroid has an elongated, high-inclination orbit which would classify it as an intermediate-period comet except that it shows no signs of outgassing, seeming to be totally inert. Its name was chosen to remind us of the story about the Sword of Damocles, because its future orbit has a good chance of evolving into an Earth-crossing one, although presently it comes no closer than Mars. If this asteroid should hit the Earth, the consequences would be horrendous, Damocles being about 20 kilometers in size.

Damocles is best thought of as a comet which is extinct or dormant, whereas an example of a dormant comet that has recently "woken up" is P/Encke (see Figure 3). This object was discovered toward the end of the eighteenth century and is quite bright, returning every three years or so (that in itself being indicative of an orbit considered basically asteroidal). The only plausible explanation as to why it was not seen previously seems to be that it earlier had no significant outgassing and so was unobservable using the instrumentation available two centuries or more ago; that is, prior to the late eighteenth century, it would have appeared asteroidal if the necessary equipment had been available to make it detectable. If it had been as bright earlier as it is now, it would have been discovered with the naked eye. In addition, there are several other comets with very short orbital periods that have been observed on one passage only and not again despite being sought, perhaps due to an insulating crust choking their outgassing and thus making them much dimmer in appearance. The point is that a bucket full of water is easily seen if it has been vaporized into a huge cloud, but not if the water remains frozen in the bucket, especially if that receptacle is black.

Back, then, to our determination of the total populations in each class shown in Table 1, bearing in mind that this classification is not definitive and is becoming hazier as the years go by. This determination may seem to be a daunting task—if we know of only a small fraction of the total, how can that total be calculated? In reality, however, it is quite straightforward. All that needs to be done is to ask, "How often are known objects accidentally rediscovered?" If there were a handful of teams searching the skies, and 90% of the time when they spotted an Earth-crossing asteroid it was later realized that the object in question was previously known, clearly it would follow that the majority of those objects had been identified and tracked. However, if only 10% of the time their apparent "discoveries" turned out to be known targets, with the other 90% being objects that had not been observed previously, then obviously only a small fraction of the total population of Earth-crossing asteroids would have been discovered.

In fact, the completeness of discovery depends on the size of object that is determined to be pertinent, which means the faintest objects that can be detected with the particular search telescopes. It is believed that no near-Earth asteroids larger than about 14 kilometers in size await discovery, because such objects would be detectable at large distances from the Earth. For stoney-type asteroids, which are brighter than the less common carbonaceous asteroids, it is believed that all such objects larger than 7 kilometers have now been found. There are only a handful of Earth-approachers with such dimensions, such as 2212 Hephaistos (which is an Earth-crosser, an Apollo) and 1627 Ivar and 1580 Betulia (which both have Amor-type orbits, but may later evolve to become Apollos). Note that the low numbering of these objects (the sequence now exceeds 6,000), is indicative of the fact that most of the very large near-Earth asteroids were found some time ago. There are occasional surprises, though; we found 1992 HE from Australia recently, that object being around 5 kilometers in size. (See the Glossary for details of how asteroids are named.)

However, there are few objects as large as these, either asteroids or comets on Earth-crossing paths, meaning that they hit the Earth very infrequently. As explained in more detail later in the book, the size threshold at which a global catastrophe would be provoked by an impact is somewhere between 1 and 2 kilometers, so what we are immediately interested in is the number of asteroids and comets that are larger than about a kilometer in size. So, what can we say about how many Earth-menacing asteroids and comets of 1-kilometer size exist?

At the time of this writing, 160 or so Earth-crossing asteroids are known, although many have poorly determined orbits and have been lost again. Of these, about half are more than 1 kilometer across. Photographic-based searches for Earth-approaching asteroids have been performed for a couple of decades, most of the finds from those programs being larger than 1 kilometer; any smaller asteroid would have to be in our celestial backyard to stand an appreciable chance of detection. It has only been in the past few years that a telescopic camera at the University of Arizona, known as Spacewatch, has been operated. It is able to detect 10- to 100-meter asteroids as they pass near the Earth. For most search programs, even a 1-kilometer Earth-crosser is unlikely to be found unless it happens to be at the appropriate place in its orbit when the observations take place. According to best estimates, about 15% of the total population of the Earth-crossing asteroids down to 2 to 4 kilometers in size is known, and about 7% of the 1- to 2-kilometer bodies. Using these percentages, one can arrive at an estimate of about 2,000 Earth-crossing asteroids larger than the minimal global catastrophe threshold, a diameter of 1 kilometer.

So how often is an impact by one of these asteroids—Apollos and Atens—to be expected? Using the data in Table 1, their mean collision probability is about 15 per billion orbits, and we multiply that figure by 2,000 and divide by the mean orbital period of about 2 years, to derive an estimate of the impact rate: about once per 70,000 years. Because the larger (bigger than 1 kilometer) asteroids tend to have lower collision probabilities than the overall known population (because some of those contributing to Table 1 are small asteroids in Earth-like orbits discovered by the Spacewatch team), a figure of once per 100,000 years is a useful bench mark to keep in mind for the larger bodies. It is important to note that this estimate could well be wrong by a factor of a few in either direction. For example, the number of Aten asteroids is highly uncertain, and yet these have relatively high impact probabilities per *orbit* (Table 1); their very short orbital periods mean that they have exceptionally high impact probabilities per *year* (large values of P/T). There may be a couple of hundred Atens, or there might be many more. Those with aphelia just beyond 1 AU are the most difficult to spot, but they are also the most dangerous ones because they have extreme terrestrial impact probabilities.

What of the smaller Earth-crossing asteroids? The pebbles one picks up on the beach follow what is known as a *power-law distribution*, with the number at any size perhaps being proportional to the inverse of the square of their sizes. The results of calculations based on this law lead to the following being the estimated populations of smaller Earth-crossers:

Larger than 1 kilometer: 2,000 (maybe 1,000 to 4,000)

Larger than 500 meters: 10,000 (maybe 5,000 to 20,000)

Larger than 100 meters: 300,000 (maybe 150,000 to 1 million)

Larger than 10 meters: 150 million (maybe 10 to 1,000 million)

Coupled with the collision probabilities of Table 1, impact rates for different-sized asteroids can be determined, as shown in Figure 5.

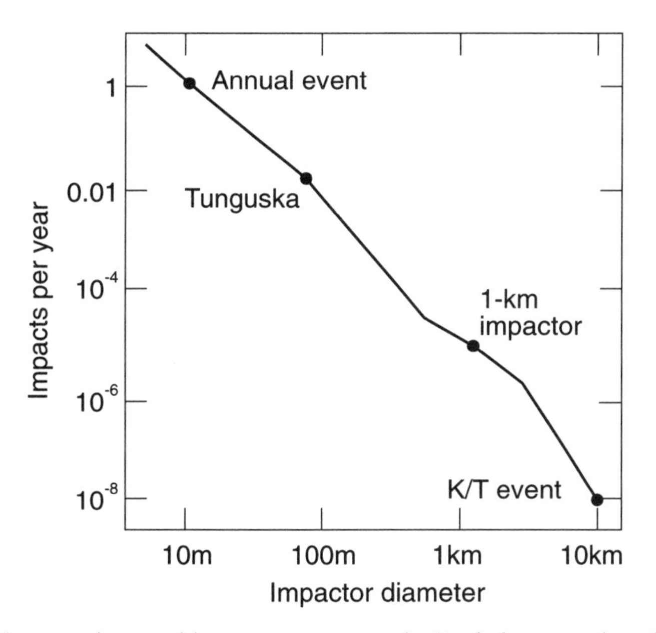

Figure 5. The typical interval between impacts on the Earth for asteroids and comets of different sizes. An event such as the Tunguska explosion may be expected about every 50 to 100 years. The annual event is in the 20- to 50-kiloton energy range, but still highly uncertain: There is still much argument over the interpretation of the rudimentary data in hand. At the higher energy end of the spectrum are 1-kilometer asteroids releasing 100,000 megatons (10^{11} tons) of energy that impact about once every 100,000 years. There is, again, uncertainty at the very largest energies because very few objects capable of delivering such a blow are known, but a best guess makes impacts such as the K/T boundary event a once every 50- to 100-million year occurrence. The K/T event (when the dinosaurs died) is described later.

As for objects smaller than 10 meters in size, our best calculations estimate that due to their greater number, they will enter the Earth's atmosphere at a rate of approximately one per year. They are of little concern, however, because most will detonate in the atmosphere and not reach the ground. These bodies are observed as very bright fireballs. Indeed, you would be exceedingly unfortunate to be hit by a meteorite or to suffer any other ill consequences from such small Earth-crossing asteroids. The chance of being hit by some part of a jetliner, such as an engine falling off of a wing or waste material being dumped, is much higher.

Later in the book, we will see why the atmosphere acts as such an effective barrier to most impactors of 10 meters or less in diameter. The atmosphere is not as impenetrable to the larger impactors, however. An object that entered the atmosphere above Siberia in 1908 was about 60 meters in size, and the blast wave flattened thousands of square kilometers of forest. How often would such events be expected to occur? The answer, again using the probabilities from Table 1, is about once every 50 years if there are a few million similar objects in Earth-crossing orbits. Factoring together the populations and impact probabilities for different types of objects, the annual terrestrial influx can be calculated, as shown in Figure 6. Of course, the influx of meteoroids and dust is determined directly and thus is relatively well known; the influx of larger objects is rather more uncertain.

The attentive reader will have noted that 1908 was over 80 years ago, and events like the Siberia explosion, known as the Tunguska explosion, are stated here to occur perhaps as often as every 50 years. This may give the impression that we are "overdue" for another one,

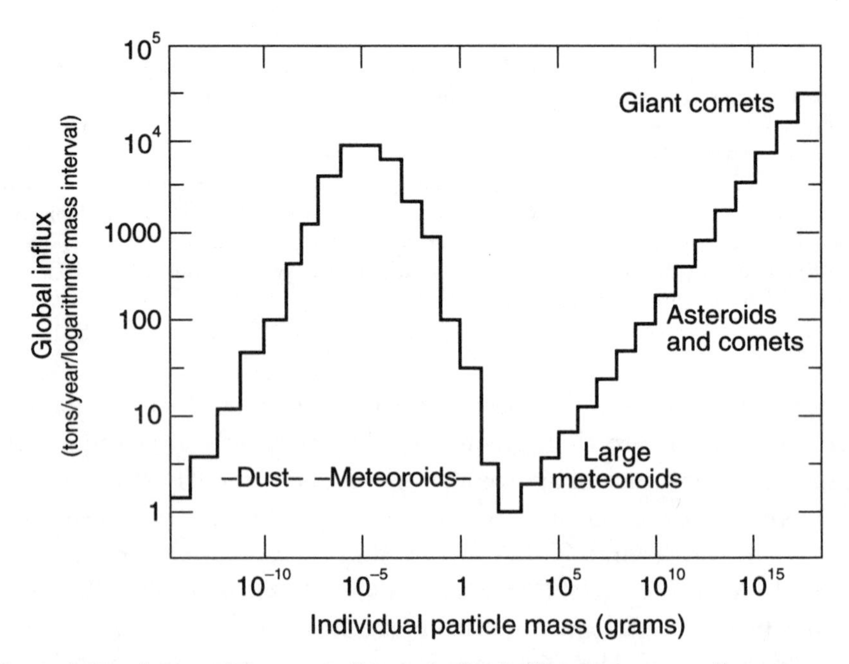

Figure 6. Distribution of the mass influx to the Earth. The mass continually arriving on the Earth as small particles (meteoroids and dust) is about 40,000 tons per year, around 20% to 30% of the total. The long-term averaged mass influx from the very largest objects is very uncertain. Note that giant comets (or impacts like that at the K/T boundary) may be expected to dominate the long-term mass influx, even though they occur very infrequently.

but even the most ardent campaigner for more research on these objects would not stoop to saying such a thing. In a Poissonian process (one in which events occur randomly in time), one is never "overdue" for an event. In this case, we might go a millennium without having a 50-meter asteroid hit the Earth (although it is very unlikely), and then have two arrive within a few days of each other (which is similarly unlikely). It would not be unusual, though, to have a century or two elapse between such impacts, and then have two within a decade. To turn things around, people often ask, "How long might we expect to wait until the next impact?"—to which the answer is, "No time at all": For a Poissonian process, the most likely waiting time is zero. In the same vein, it should not be imagined that a recent event provides any protection against future events. For example, there was much angst in Great Britain when in the late 1980s a ferocious storm uprooted many centuries-old trees and caused widespread property damage. The meteorologists stated that this was a once-per-200-year storm, so there was a public outcry when a similar storm occurred a couple of years later. The error, however, was in the public's interpretation of the climatologists' statements. Future events are not affected by past events: It is not like when you just miss your bus or train, and you know you have to wait another hour until the next one is due. Poissonian processes do not follow timetables. Note also that throughout this book I make statements along the lines of "such an impact occurs once every 100,000 years." What is meant is that the average time between impacts is 100,000 years. This is *not* intended to imply that impacts occur as regularly as clockwork every 100 millennia. They don't, because they are not like buses.

The chances of an impact by a comet have not yet been considered, and this is because the calculation of these is very problematic. On the one hand, there is little doubt that the very largest possible impactors are comets and that these on average have high impact speeds, so that it is comets that pose the major threat of truly cataclysmic impacts, perhaps causing mass extinctions. On the other hand, we are woefully ignorant of the masses of comets, which means that it is exceedingly difficult to make any meaningful estimate of their contribution to the impact hazard. It is usual to quantify matters by talking about the contributions by different classes (that is, those listed in Table 1) to the impact events on Earth that produce craters larger than 10 or 20 kilometers in size. This bench mark encompasses the combined influences of mass and impact speed, and one notes from Table 1 that cometary impact speeds tend to be higher than those of asteroids. Recent estimates for the cometary contribution have varied between 2% and 30% of the total, a fifteenfold difference. In addition, many of the Earth-crossing asteroids—perhaps 50% or more—are extinct or dormant comets; does one include these in the cometary contribution to the impacts, or not? Here, such objects have been included with the asteroids.

The major source of the uncertainty is, however, our ignorance with regard to how to relate cometary brightnesses to their nuclear masses. A comet that has only 1% of its nucleus actively losing gas into the coma (the other 99% being choked with silicates and heavy organic chemicals) would not be expected to be as bright as one with 20% of its surface being active; but then the former might be rich in very light volatiles and so lose a great deal of gas into its coma from that small active area. The activity will also vary with distance from the Sun, and many comets are not symmetric in their activity about perihelion. Even if we understood a "standard" comet, it would not mean that we could then quantify the behavior of other comets; in any case, as we have seen, we are uncertain of the mass of even our best candidate as a standard comet (P/Halley, observed for more than 2,000 years and visited by five spacecraft in 1986) by a factor of 3.

Even within the comets, it is difficult to allot reasonable estimates of the separate contributions of short-period, intermediate-period, and parabolic comets. The distinctions are important. Because the short-period comets come back frequently, many of them have been subjected to detailed study over the past 20 years. However, it is likely that our census of these objects down to 1-kilometer sizes is far from complete, with many more yet to be found. This group is often called the *Jupiter family*, and most have aphelia close to Jupiter's orbit, with that planet likely to have been responsible for capturing them into their present paths from much larger orbits. Because these short-period comets are all prograde, they make frequent (astronomically speaking), low-velocity, close approaches to Jupiter and their orbits can alter quite rapidly. In general, they are unstable on timescales of 10,000 years, but notable changes in their orbits can occur over mere centuries or millennia. The comets used in Table 1 include only those that are Earth-crossing: There are about ten times as many known short-period comets that have perihelia further out in the solar system. A large fraction of those have perihelia between 1 and 2 AU, so that quite small jovian perturbations can lead to them attaining Earth-crossing orbits. The values derived from our present "snapshot" of the population is therefore not necessarily a good measure of the long-term average.

Although only 10% of the known short-period comets are Earth-crossing, the 13 intermediate-period comets included in Table 1 represent more than 50% of the total of such bodies. This does not, however, necessarily mean that of all the intermediate-period comets (including those as yet undiscovered) half or more are potential Earth-impactors. Many unknown intermediate-period comets have yet to return to perihelion since the invention of photography, so those that are intrinsically faint would not have been discovered in the past and their first chance at discovery is yet to come. The intermediate-period comets more likely to have been discovered since antiquity are those that have perihelia within 1 AU because (1) they may come closer to the Earth; and (2) they would be intrinsically brighter because the intense solar radiation causes the evaporation of more volatiles, and therefore a larger, brighter coma and tail. Another line of reasoning says that the presence of two intermediate-period comets larger than 10 kilometers in size (P/Halley and P/Swift-Tuttle) leads to an expectation of at least 100 intermediate-period comets larger than 1 kilometer in size on Earth-crossing orbits, if they follow a size distribution similar to the asteroids.

Should anyone doubt that there are many comets yet to be found, they should note the following. When comets are discovered, they are given a file code that consists of the year and a lowercase letter (for example, 1992t was given to P/Swift-Tuttle). In 1985, at the triennial General Assembly of the International Astronomical Union, Brian Marsden (who keeps tabs on such things) said that it was important that the IAU made a decision as to how comets were to be labeled when more than 26 were discovered in any year (that is, when all letters in the alphabet are used). Many of the astronomers present were scornful of this suggestion because they believed that all comets, or nearly all, were being discovered. Marsden was insistent, however, and it was decided that the format 1994a_1, 1994b_1, and so on, should be used for the twenty-seventh, and subsequent comets, should the need arise; other possibilities, such as capital letters, are ruled out by their prior usage for other astronomical phenomena. Two years after his suggestion, Marsden was proven correct, with the alphabet being "overflowed" for the first time in 1987. Considering, then, this leap in comet discovery rates over the past decade, it is clear that there are many intermediate-period comets that await discovery because they have not returned to perihelion since humankind has had the equipment and the vigilance to spot them. Such has been the proliferation in comet discoveries that this designation system is being changed again in 1995.

All things considered and best estimates made, it seems likely that short-period comets and intermediate-period comets contribute something like 5% of the 1-kilometer-plus impactor flux to the Earth. This figure, which is often debated, will undoubtedly be amended as time goes by and we understand more about asteroids and comets, but it seems clear that periodic comets pose a non-negligible but minor fraction of the large impact hazard.

Parabolic comets pose their own problems in a number of ways. Because many of these are passing through the planetary region for the first and last time (having been given a nudge from the planets such that they are not expected to return), it is inappropriate to give them a characteristic period in order that an impact rate may be calculated, as was done for the other objects. Rather, it is appropriate to try to determine the flux (that is, the number per year or per millennium) of parabolic comets above some limiting mass that are on Earth-crossing orbits. The problem with this idea is that their masses are even more uncertain than those of the other comets. The reason for this is that "new" parabolic comets tend to be especially bright: Those passing through the inner solar system for the first time will be outgassing their high-volatility constituents (such as methane and carbon monoxide) at a phenomenal rate, whereas comets that have traveled through many times before will have long-since lost those constituents. On the other hand, there is no apparent minimum size limit for parabolic comets because, unlike the case of the other comets, total de-volatilization will not have occurred. On top of that, the mean impact speed of these comets is high (Table 1). Overall, it seems that parabolic comets may compose 25% of the impact hazard, but it would be a fool who believed that we know for sure that contribution to within a factor of 2 or 3.

Another problem with parabolic comets is that any future impact is unpredictable. For all of the shorter-period objects, the idea is that they are discovered and their orbits are accurately determined, so that any impacts on later orbits are predictable, providing warning times of some years and possibly decades. However, parabolic comets come our way once only, and any such impactor would be discovered on the orbit in which it is going to hit the Earth, with minimal warning—possibly as much as six months, but likely rather less, with present search activities.

The Spaceguard plan would lead to essentially all Earth-crossing asteroids and short-period comets larger than 1 kilometer being discovered and their orbits determined in a 20- to 25-year program. Obviously, it would not be possible for all intermediate-period comets to be found in that time, because their 20- to 200-year periods mean that perhaps only 50% would return during the Spaceguard survey. However, intermediate-period comets pose only a small fraction of the danger to humankind, so that by the end of the project it would be possible to say with at least 95% (and likely more than 99%) confidence whether an impact by any large periodic object was anticipated within the next two centuries. But what about smaller projectiles?

Have you ever seen an advertisement on TV for automobile insurance in which some unlucky motorist's car slips its parking brake and rolls down a hill, where it happens to scrape along the side of a Ferrari and a Rolls Royce, causes another driver to swerve into someone's swimming pool, and then topples a power pole through the roof of a mansion before speeding up a ramp into the back of a furniture truck that is being loaded with priceless antiques? Of course, the insurance company's message is that you are covered for such catastrophes by their policies.

The point is that the once-every-100,000 years 1-kilometer asteroid impact is also close to the worst possible scenario. Just as the insurance company knows that it will rarely, if ever, have to pay out on an accident such as that portrayed in its advertisement (with most claims being for smashed headlamps, dented fenders, and other minor damage caused by humdrum scrapes), so it goes for cosmic impacts on the Earth.

By far the most likely impact to occur during our lifetimes is another event like the one in 1908 when a 50-meter chunk of rock blew up in the atmosphere above Siberia. The small likelihood of a truly global devastation happening in the near future should not alleviate all our worries, however, because the effects of the much more likely impact of a smaller body might well still be catastrophic for civilization. To get a clear sense of exactly what the consequences of such an impact would be on the Earth, we will discuss plausible scenarios in the next chapter.

3

Catastrophic Scenarios

A 500-meter asteroid crashing into a desert—for example, the Outback of Australia or the Sahara—would devastate an area of about 160,000 square kilometers and cause substantial damage over a far greater region. For example, if the asteroid fell in the Outback, all of the cities in Australia would almost surely be shaken flat. This would not, however, be the worst-case scenario. The consequences of the same asteroid arriving a few hours earlier and perhaps landing in the Pacific Ocean between New Zealand and Tahiti would be far worse. This is because the impact would generate an enormous *tsunami* (a huge ocean wave), often caused by earthquakes. Over the past two centuries, Japan has been hit by many devastating tsunamis, which is why the name for such waves comes from the Japanese word.[1] These tsunamis have resulted from earthquakes as far away as South America, propagating all the way across the Pacific Ocean, causing major damage to Japan's coastal cities as the huge momentum behind them is released as they reach the continental shelf at the edge of the Japanese coast. At times, the Hawaiian Islands and other Pacific islands have also suffered awful consequences from tsunamis, but these islands have most often been protected by the fact that they stand on the peaks of sub-ocean volcanic pinnacles, so that the tsunami can wrap around them rather than break, as they do when they meet the continental shelf in the northwestern Pacific Ocean.

A historically recent severe tsunami is that triggered by an earthquake off the coast of Chile in May 1960, which killed at least 1,000 people there. About 14 hours later, it reached Hawaii, just over 10,000 kilometers away. Although, at such a range, the "deep-water wave" (referring to the wave prior to its running up to the land mass) was small—the actual tsunami height in Hawaii was between 10 and 15 meters—61 people were killed in Hilo harbor. In Japan, which the tsunami reached several hours later, the deep-water wave was only 0.2 meters high, but it produced a tsunami varying between 1 and 5 meters in height, depending on the local topography. More than 200 people met their deaths as a result.

There have been many other tsunamis in the past half century, especially in the Pacific Ocean. In 1946, a tsunami generated by an earthquake in the Aleutian Islands killed a handful of people in Alaska and more than 150 in Hawaii. Damage was inflicted on Hawaii by tsunamis produced by earthquakes in Kamchatka in 1952 and the Aleutians, again, in 1957. Another Alaskan earthquake in 1964 generated a tsunami killing 50 people in that state. A Hawaiian earthquake produced a tsunami killing a few people there in 1975, and a Japanese tremor in 1983 killed more than a hundred people in Japan and Korea. Obviously, tsunamis are not to be taken lightly, especially since much larger waves might be expected from impact-generated events.

The threat of tsunamis caused by asteroid impacts has only recently been recognized, due to the work of Jack Hills and Patrick Goda of the Los Alamos National Laboratory in New Mexico. They have performed calculations showing that the hypothetical 500-meter asteroid mentioned earlier would produce a deep-water wave 50 to 100 meters in amplitude, even at a range of 1,000 kilometers from ground zero. Since the tsunami height could be amplified by a factor of 20 or more in the run-up as continental shelves are encountered, we are referring here to a tsunami several kilometers in height. Even if the impact *were* between New Zealand and Tahiti, the tsunami breaking on Japan would be perhaps 200 to 300 meters high, and heaven help New Zealand and Tahiti themselves!

Would this simply be a matter of a huge wave hitting the coastline, with only people living or working at the beach or in the harbors being affected? There are formulas for calculating the distances inland that tsunamis propagate. For a 200- to 300-meter tsunami hitting a built-up city area (in which the buildings provide an impediment to the wave, dissipating its energy), the tsunami would travel between 50 and 100 kilometers inland, and further across flat areas. Anyone who has taken the Shinkansen (Bullet Train) between Tokyo and Osaka knows that a large fraction of Japan's people live along the coastal strips and would have little hope of survival in the case of an asteroid impact anywhere in the Pacific region (except that maybe 12 hours' warning might be available for evacuation into the mountains). The American reader might note that the same would apply to all people living within an hours' drive of the coast from San Diego to Seattle and Vancouver; not that east coast dwellers should feel smug, because the Atlantic region also presents a very large target, with cities on both sides of the Atlantic Ocean certain to be devastated by an asteroid impact anywhere between New York and London.

Is there any evidence of such giant tsunamis in the past? Obviously, now that scientists have been alerted to the possibility of impact-produced tsunamis, searching will be done, and we know what questions to ask. Already, however, coral has been found several hundred meters above sea level throughout the Hawaiian Islands, some as high as 326 meters on Lanai. Although the precise way that the coral was transported to such a height has not been established, it seems clear that giant impact-driven tsunamis could well have been responsible.

This terrifying scenario raises a central point about how we should consider the gravity of, and the risk involved in, the various catastrophic scenarios that might be caused by an impact. When we consider these scenarios, which we will do in this chapter, there is probably always one central question in each reader's mind: What is the chance of my dying as the result of an asteroid or comet impact? Though asking this makes good sense, and is the advisable way of evaluating the dangers posed, we must be aware that the issue of how to calculate these chances not only is contentious (in that there are differing opinions as to how frequently the Earth is struck), but also is one about which there is a great deal of uncertainty in the scientific sense. While the public often believes that scientists are able to put hard and fast numbers on everything, in fact, in many areas it is just not possible to be sure of the consequences of any particular event, because certain things, such as the response of the Earth's complex ecosystem to a major insult, have so many variable parameters that it is beyond science to derive a theoretical solution with 100% certainty. This is often the case in the real world, with apparently simple situations actually being intractable.

As someone born in England, I like to drink tea poured from a china or claywork teapot, and in order to pour properly (without a dribble), the nozzle of the teapot must be properly designed. This is a problem of fluid mechanics, but it is not one that is solvable by methods of theoretical physics. Teapots have spouts that pour properly due to many years of experimentation by artisans; that is, teapot design is an art, not a science. In the same way, it is still necessary to subject models and prototypes of airplanes and boats to tests in wind tunnels and artificial ponds. Such experiments—for ships, planes, and teapots—can also be done by computer using numerical simulations, but these are just numerical experiments simulating the flow, not analytical solutions, and are achieved by brute force, not finesse.

For these problems of fluid dynamics, then, the remedy is to do an experiment. Cosmic impacts constitute a phenomenon with which we should be loathe to dabble experimentally, however: We want to avoid asteroid impacts upon the Earth, not cause them. We might hope that we will be able to watch impacts on other planets from time to time and learn from our observations, the collisions of the fragments of a comet with Jupiter in July 1994 being a wonderful case in point. But, in the main, we need to construct theoretical models of how the terrestrial environment might respond to such impacts, perhaps with geological evidence from past events to help corroborate the results derived. A discussion of these models is essential to our understanding of the various scenarios that have been developed.

Foremost in providing estimates of the relative risk to individuals of asteroid and comet impacts have been David Morrison, of NASA–Ames Research Center in California, and Clark Chapman, of the Planetary Science Institute in Tucson, Arizona. They have tried to help the public understand their predictions by comparing the risks they have calculated to the risks of other ways in which people meet their demise in accidents. I differ from their interpretation somewhat in my evaluation of the impact hazard, but broadly agree with their analysis pertaining to large, 1-kilometer-plus impacts. All in all, I think their analysis provides a useful means of studying how impactors interact with our planet, and hence what hazard they pose. Next we will examine how that hazard can be quantified.

As previously mentioned, our atmosphere provides an efficient shield against incoming asteroids and comets up to a certain size. That size depends on a number of parameters, especially the speed, entry angle, composition, density, and physical strength of the object, but as a rule of thumb, 50 meters is the smallest size of rocky asteroid that might reach the surface largely intact, and 100 meters the smallest icy cometary object. A very rough guide indicating the minimum size of object that might be expected to reach the surface can be gained from the following simple reasoning. When watching the weather report on TV, the pressure in different locations is usually quoted in millibars, with 1013 millibars (or 1.013 bars) being the *standard atmospheric pressure*; this is equivalent to about 14 pounds per square inch (psi), which is the normal unit used for tire pressures. In metric units, a bar is a pressure of 100,000 Newtons per square meter (100,000 Pascals, in another unit). A Newton is a unit of force, which (strictly speaking) gives the weight of any object; to convert from a force in Newtons to a mass in kilograms, we divide by the acceleration due to gravity, 9.81 meters per second-squared. The atmospheric pressure of 100,000 Newtons per square meter means, therefore, that the total mass of gas above each square meter of ground at sea level is just over 10,000 kilograms (or 10 tons). We now make the intellectual leap of deciding that any incoming object might reasonably be expected to reach the ground intact if, in its passage through the atmosphere, it meets less than its own mass of gas. This means, then, that the minimum size would be such that a one-square-meter section through the body has a mass of 10 tons. If its density were the same as that of water, this would mean a section 10 meters long, and if the density were like that of rock, about 4 meters long. However, these figures are for an incoming object that would need to be cube-shaped; for a spherical shape, the diameters would be a little bigger, about 12 and 5 meters, respectively. In addition, in this example, we effectively assume that the objects are entering vertically, whereas the most likely angle of entry is 45° to the vertical, in which case a larger mass of atmospheric gas is encountered. The bottom line is that an icy projectile with a size less than about 20 meters might be expected to survive to the ground if it has a very low entry speed and similarly a rocky body smaller than about 8 meters. At higher speeds, an object needs to be larger to survive entry through the atmosphere. For example, if you do a belly flop off of the side of a swimming pool, it's not too bad, but if you do the same thing off of the top diving board, it hurts! More detailed analyses imply minimum sizes of about 50 meters for rocky bodies and 100 meters for icy ones. Smaller metallic asteroids can penetrate the atmosphere, and such objects are known to exist, but these comprise, it is believed, at most 3% of the impactor flux.[2]

The vast majority of small asteroids and comets, then, do not reach the ground. This is not to say, however, that such objects are not dangerous, as the example of the damage caused by the asteroid that blew up above the Tunguska River region of Siberia in 1908 makes clear. In this case there is no impact crater and no substantial residue from the incoming object. This is despite the fact that its explosive energy was similar to that of the impactor that formed Meteor Crater; the difference is that the Tunguska object seems to have been of rocky composition, while the Meteor Crater excavator was nickel-iron. Studies of the damage caused by nuclear weapons show that an airburst causes more destruction than a groundburst (to which extent it is preferable in many ways that a cosmic impactor reaches the ground rather than detonates some kilometers up), and also that the area of ground laid to waste varies as the energy of the explosion to the power 2/3 (that is, the square of the cube root; for example, the 2/3 power of 10 is 4.64). Chapman and Morrison point out that the Tunguska explosion (which occurred at an altitude of between 6 and 10 kilometers) provides a calibration point for such events. The devastated area of forest was about 2,200 square kilometers, and the energy has been estimated as being between 10 and 20 Mt. A rough estimate of the expected area of devastation is therefore given by

$$A = 400\ E^x$$

where $x = 2/3$, A is in square kilometers, and E is the explosive yield in Mt. Applying this formula to a million-Mt explosion, one derives an area of 4 million square kilometers, which is half the area of the continental United States. Such an incoming object would, of course, reach the ground and form a crater roughly 20 to 30 kilometers in diameter. The ejecta would cause damage over a much wider area—in fact, globally.

Events like Tunguska occur with uncertain frequency, possibly once every 50 years, if the interpretation of the Spacewatch data is correct, or at most once every 300 to 500 years. ('Spacewatch' is a University of Arizona telescope that we will describe later.) Moving up a little in size, the evidence of lunar craters agrees with the flux of objects larger than 100 meters detected by Spacewatch, so here we appear to be on firm ground, figuratively speaking. For a 100-meter object, the impact energy (at a speed of 22 km/sec) would be about 100 Mt, with the devastated area about 10,000 square kilometers. This is twice the area of Delaware, about the same size as that of Connecticut, and about half the area of Massachusetts or New Jersey. How often is such an event to be expected? We believe that it would be about once every millennium, meaning that during the normal lifetime of a person in a western country there is a 7% to 8% chance that such a catastrophe will occur *somewhere*. The overwhelming odds, of course, are that if this did happen over land, it would be a sparsely populated region; then again, the chance is about 70% that it would occur over the ocean, and that would most likely spell disaster for many coastal cities because of the tsunami produced. Shin Yabushita, of the University of Kyoto, who has good reasons for interest in the topic since Japan is especially at risk, has calculated that there is at least a 1% chance that practically all of the cities around the Pacific rim will be obliterated within the next century by an asteroid-induced tsunami.

What might be the expected number of deaths from an event such as a 100-meter asteroid's arrival? The average population density on the land area of the Earth is about 30 per square kilometer, varying from about 5,000 in Hong Kong and the District of Columbia down to 1.5 per square kilometer in Australia and just 0.1 in Alaska. Because two thirds of the Earth's surface is covered by water, the global average is ten people per square kilometer. Therefore a 100-meter impactor, detonating in the atmosphere and laying waste to 10,000 square kilometers of land mass, might be expected to kill 100,000 people, neglecting the effects of a tsunami. Any estimate of the effect of a tsunami is still debatable, but the ocean's ability to transport the energy of the impactor through the wave over a long range to the very regions where human settlements tend to be concentrated (coastal plains) warrants at least a tenfold increase in the expectation of deaths in addition to that from direct blast, possibly over a hundredfold. That is, for small impactors, we profit from the fact that humankind is concentrated in certain areas, the atmospheric detonation most likely being over an unpopulated region, but we lose from the fact that the oceans are able to propagate the effects of such an explosion over a much wider area, perhaps a whole hemisphere. We are therefore talking about minimally a 1-million-death event (100,000 in the direct blast, the rest because of the tsunami) every thousand years, and a 1 in 10 to 15 chance that any particular person will witness such a calamity somewhere on the Earth during their lifetime. In order to provide some quantization for comparison with larger impacts, note that this implies an average of about 1,000 deaths per annum, although that figure is highly uncertain and might be a lot higher.

To illustrate the difficulty of such calculations, consider the Pacific Ocean as a target. This ocean covers 25% to 30% of the Earth's surface, so we expect an impact by a 100-meter asteroid over the Pacific about every 4,000 years. Such an event would produce a tsunami that could deluge all of the cities around the Pacific rim. Considering only Sydney, with a population of about 4 million, if these inhabitants were all killed, the annual death rate averages out to 1,000 a year for Sydney alone. When the other cities are included (Los Angeles, San Francisco, Seattle, Vancouver, Tokyo, Shanghai, Hong Kong, Taipei, Manila, Auckland, Brisbane, and so on), clearly the death rate calculated is much higher, and it draws into question whether the San Andreas fault is the most significant natural hazard to which a resident of San Francisco is exposed. In fact, more than 300 million people live in the Pacific region, and it is interesting to note that although humankind had spread into the large islands of Australia and New Guinea by about 40,000 years ago, it was not until the last 5,000 years that the smaller Pacific islands have been populated, according to available archaeological evidence. We might therefore speculate whether past impacts produced tsunamis that swept the Pacific islands clear of humans in some earlier period.

We now move on to the much larger impacts that occur much less frequently. For the sake of argument, we consider the million megaton explosion (due to a 2-kilometer asteroid and expected about once every 500,000 years) and also the same sort of impact that killed off the dinosaurs, a 100-million megaton explosion (due to a 10-kilometer asteroid or comet). Even the lower of these energies is almost certainly in excess of the minimum required for a global catastrophe, the threshold being in the 1- to 2-kilometer range, according to the most authoritative models. But first we must address the question of what exactly is meant by *global effects*.

Clearly the object(s) that hit the Earth at the time of the death of the dinosaurs 65 million years ago caused global effects: Anomalous quantities of rare metals and isotopic anomalies have been found in the relevant stratum worldwide, as has an excess of certain extraterrestrial amino acids and soot in quantities indicative of at least 90% of the global biomass being incinerated. In 1980, the Alvarez team, working back from the amount of iridium (one of the rare metals) that they had identified, deduced an impactor size of at least 10 kilometers,but the size of the crater recognized in the Yucatán in 1990 (at least 180 kilometers, perhaps 400 kilometers, in diameter), indicates that the projectile might have been closer to 20 to 40 kilometers in size. The frequency of such impacts cannot be estimated with any confidence, but it is likely to be in the once per 10- to 100-million-year range. The reason for the uncertainty is that we have insufficient data. There appear to be no Earth-crossing asteroids as large as 10 kilometers in the present epoch, but that does not mean there were no such asteroids in the past. Such bodies may arrive in Earth-crossing orbits in the future. Among the comets, unless we are wildly wrong with regard to the sizes of cometary nuclei, it is only P/Halley and P/Swift-Tuttle that are of these dimensions among the Earth-crossing periodic comets, giving a sample too small to provide a sensible estimate of the terrestrial influx. Of the parabolic comets, some are certainly of at least 10-kilometers dimensions and greater, but these are seen so infrequently that over historic times too small a number have been observed to allow a reasonable estimate of their flux. Lunar craters give us our best measure of the flux of 10-kilometers-plus near-Earth objects, but even then the numbers involved (along with the uncertainty in the speed and nature of the impactor in each case) are too small to allow any definitive answer.

Even if the rate of such enormous impacts is as great as one per 10 million years, for a world population of about 6 billion people, this implies an annual expectancy of deaths of at most 600, which is greatly exceeded by the 100-meter objects considered earlier. This is not to say that the annual expected mortality rate is the most significant parameter. There is an important qualitative difference between an impact killing tens of thousands of people, lamentable though that would be, and an impact that could wipe out the human race. Evolution has built into us, and all species, a conscious, and often unconscious, desire to propagate our genes. To the extent that all of humankind shares genetic material, an impact causing the deaths of even 99% of the human population is therefore an entirely different proposition to that which reaches the critical 100% level. Nevertheless, we are considering here the economics of Armageddon and are quantifying our possible responses in terms of mean annual death rates for different classes of impactor size. So far, we have deduced that 100-meter projectiles may be expected to produce more deaths, when time-averaged over many millions of years, than the much more powerful 10-kilometer impacts, because the latter are very infrequent. It's like hunting bison: If the first people to bring guns into the American West had not almost wiped out the species with their profligate shooting, then a greater number of bison would have been killed in succeeding decades, but many more than now exist would still roam the prairies.

What about the 2-kilometer impactor? The idea of "Nuclear Winter"—in which the global wildfires following a major nuclear war inject so much soot into high altitudes (taking some weeks or months to settle out) that the surface temperature of the Earth is cooled to such a degree that an entire agricultural season is lost—was itself prompted by studies of the climatic effects of the Tunguska event. Various possible climatic repercussions from a nuclear war have been studied, such as the initiation of midsummer frosts that would prevent the cultivation of many types of crop, shortening of the growing season for many others so as to make ripening impossible, environmental catastrophes wreaked by ozone depletion, nitrogen oxide poisoning of the atmosphere, and the effects of the fires themselves. These consequences are similar to those that would be expected after a large asteroid or comet strike, producing a "Cosmic Winter," the only mitigating factor being the lack of radioactive fallout.

The Nuclear Winter scenarios, then, are directly applicable to the concept of a Cosmic Winter, with the critical threshold being that at which sufficient dust (micron sized and smaller) is raised up into and above the stratosphere and reflects away enough sunlight so that the surface temperature drops by several degrees. Such dust would take months to years to settle out, leading to a total collapse of agricultural production and resulting in mass starvation. The threshold energy for such an event is in the 0.1- to 1.0-million Mt range, meaning a 1- to 2-kilometer size range for an impact at 20 to 25 km/sec.

As an absolute minimum we might expect 25% of the human race to die as a result of such an event, with a more likely figure being well in excess of 50% since it would be the low-technology, mass-population, Third World countries that would be less able to cope. They have neither the advanced agricultural capabilities nor the food stores to survive through a period of duress; witness the famines that occur in Africa during every drought. Even with a figure as low as 25%, this implies 1.5 billion deaths in an event that has a probability of occurrence of between one in 100,000 and one in 500,000 per annum (depending on that uncertain energy threshold). The annual expectancy of deaths is therefore in the 3,000 to 15,000 range, which is larger than the values derived for 100-meter or 10-kilometer impactors. Thus the response of the Earth's human population in terms of the anticipated numbers of deaths among our species is nonlinear, with a peak for impactors around 1 to 2 kilometers in size.

It also turns out that this 1- to 2-kilometer limit defines the sizes that we can actually do something about. Earth-approaching objects are detectable, using presently available astronomical and military facilities, down to a certain brightness that delineates some size and range envelope. Celestial bodies smaller than about 1 kilometer in size pose a problem in that they could not be discovered and tracked within a reasonable time: They pass through that envelope too seldomly. An astronomical survey program using presently available technologies would take a half century or more of continuous operation to discover and track all near-Earth asteroids and comets larger than (say) 0.5 kilometer, because of the laws of celestial mechanics which govern how often these objects come close enough to the Earth to be observable. Many could be found within a decade, but only a fraction of the total population down to that size limit. However, the 1-kilometer-plus objects, which apparently pose the major hazard in terms of human lives lost per year due to impacts, are ripe for the picking: They could all be found in a 20- to 25-year program because they are detectable at larger geocentric distances. We know how to find them; we just need to get on with it and hope that there is not one with our number on it coming our way soon. The Spaceguard program described in Chapter 11 should begin as soon as possible; the cost is $300 million, the benefit many times higher.

This summary of the effects of an impact at the 1- to 2-kilometer threshold glosses over the way in which *global effects* are defined. Various complicated atmospheric models lead to a result indicating that sufficient dust and soot would be lofted into the stratosphere to cool the environment by a certain amount—nominally 10°C (18°F)—after such impacts, and that is sometimes used as a definition of what is meant by *global effects*. This is a definition based on physics, but surely our definition of a global catastrophe should be anthropocentric: The Earth has suffered such events many times in the past and carries on regardless. The matter of concern here is whether *humankind* would be able to carry on regardless. A definition based on the human perspective is therefore necessary. This could be argued about ad infinitum or ad nauseam, but Chapman and Morrison take the sensible step of defining what they would consider to be a globally catastrophic impact. This would be an impact that

> destroys most of a year's global food crops, and/or
>
> results in the deaths of more than 25% of humankind, and/or
>
> threatens the stability and future of modern civilization.

There is little doubt that a 1- to 2-kilometer impactor would produce these effects, the outcome being far worse than that of any past wars and comparable to the worst imaginable nuclear war.

How does the chance of dying in a impact compare with the other hazards of modern life? As stated earlier, the annual expectancy of deaths due to the threshold impact is between 3,000 and 15,000, depending on whether a 2-kilometer or only a 1-kilometer impactor were necessary to breach that threshold (a once per 500,000-year or once per 100,000-year event). Let us make the assumption that the answer is somewhere in between, and the averaged annual death rate from impacts is about 8,000 (see endnote 3). This is a bit more than a millionth of the world population, and we will assume that a similar fraction of each country might die; this is clearly a spurious assumption, but it will suffice for illustrative purposes. Considering only U.S. citizens, about 300 deaths per year are expected, whereas about 3 million deaths occur per year in the United States due to all other causes. This means that the average person in the United States has about a one in 10,000 chance of dying due to an asteroid or comet impact.

The quintessential question is, "How does this compare to other causes of death?" Most people die naturally due to old age, cardiac problems or cancers, and so on. What we need to gauge impact deaths against are the many fatalities that occur due to automobile accidents, murders, plane crashes, and so on. Chapman and Morrison have again done the groundwork here; Table 2 presents their results, except that my derived value for impact deaths (one in 10,000 chance) is twice as high as their value (one in 20,000), since I believe that they underestimated that probability. Even one in 10,000 is probably too low by a factor of a few.

Obviously, for any other country the exact figures would be different. For Australia, the top of the list would again be road accidents, and again the rate would be about one in a hundred, but other causes

Table 2
Mean Probabilities of Dying Due to Certain Accidental Causes for U.S. Residents

Cause of death	*Probability*
Automobile accident	1 in 100
Homicide	1 in 300
House or other fire	1 in 800
Accidental shooting	1 in 2,500
Electrocution	1 in 5,000
Asteroid/comet impact	1 in 10,000*
Airplane disaster	1 in 20,000
Flood	1 in 30,000
Tornado	1 in 60,000
Snake bite or insect sting	1 in 100,000
Fireworks accident	1 in 1 million
Botulism poisoning	1 in 3 million

*Value is the author's.

of death would be listed—for example, shark attacks at one in 200,000, and skin cancer toward the top of the table. I would include the latter as an act of volition because it is an easily avoided cause of death. Lung cancer caused by smoking should be similarly included.

Finally in this chapter it is appropriate to note that next to each other in Table 2 are the chances of death due to a cosmic impact and those due to an airplane crash. Surprisingly, the former is the higher, by

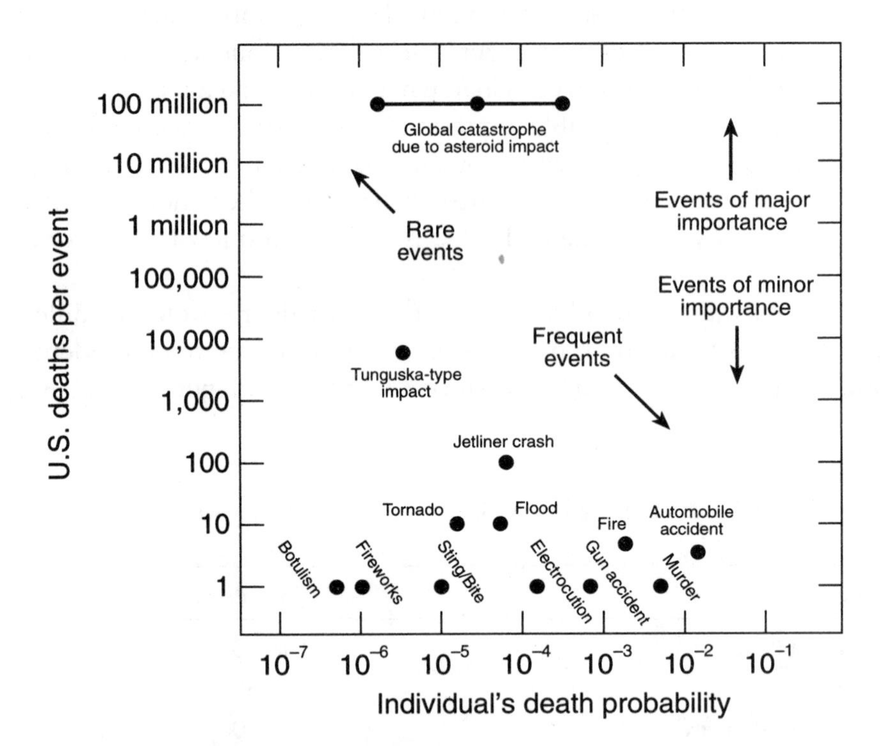

Figure 7. Public perceptions of risk do not always represent a realistic assessment of the actual hazard. Thus people tend to be worried about dying in a plane crash, because this is a high death-per-event accident (elevated on the vertical axis here), whereas automobile accidents are more likely to kill you (far to the right on the horizontal axis). Large asteroid impacts have uncertain kill-probabilities, due to our ignorance of the threshold at which a global catastrophe occurs, and this is reflected by the three possible values plotted. My own view is that the true value is toward the rightmost point (high probability) because the impact probability at any threshold is higher than generally thought. To the person on the street, major catastrophes are very worrying, and we see here that asteroid/comet impacts are the extreme example (topmost point) because they have the potential to cause a huge number of deaths, more than any other event except perhaps for an all-out global nuclear war. (After C.R. Chapman and D. Morrison, "Impacts on the Earth by asteroids and comets: Assessing the hazard," *Nature*, volume 367, pp. 33–40, 1994)

a factor of 2, but even so, many people elect not to fly. Flying is a prime example of the sort of hazard that people worry about: a small risk, but a highly emotive one. Jetliner crashes occur infrequently, but when they do, they are prone to be calamitous, with hundreds of people dying together. They are therefore newsworthy, provoking more fears among the populace about the dangers of flying. On the other hand, the media do not headline the stories about the greater numbers of people dying on the roads as they travel to the airport.

Of all mega-accidents, asteroid and comet impacts are the most extreme example: the very infrequent events which must lead to huge numbers of deaths. Some idea of this comparison between rare events and frequent events, accidents with high consequences and those with lesser consequences, can be gained by plotting the number of deaths in an accident or event against the probability of its occurrence (see Figure 7). If people reacted to the impact threat with the same approach as many do to flying, the coasts would be deserted for miles inland, and all would have fled to the high country where they would have burrowed into the side of the nearest mountain and laid in a plentiful supply of food and fuel to see their way through the Cosmic Winter, fingers crossed that the impact is not too nearby. The actual hazard does not warrant such steps, but it does warrant a much greater effort than is being applied at present. Humankind may expect between 5,000 and 10,000 people per annum to die due to impacts, averaged over many years, with a small but finite chance that billions may be killed one year soon; the total number of astronomers employed in searching the skies for the offending projectiles is about a dozen. That is entirely disproportionate.

4

The World on Fire

In the last chapter, we covered the wide spectrum of possible consequences of impacts, from the very largest, releasing energy equivalent to many millions of megatons of TNT, down to atmospheric detonations producing mainly local effects (such as the Tunguska explosion). We have not yet, however, touched on all of the catastrophic possibilities involved with impacts. This chapter will look at a number of possible effects caused by the perturbation of the Earth's atmosphere that would result after an impact, including the possibility of a global firestorm.

In the catastrophe at the end of the Cretaceous era, when the dinosaurs disappeared, a soot deposit has been found at the geological boundary that is equivalent to the combustion of at least 90% of the terrestrial biomass. The discovery of this soot layer, and its interpretation, was made by Ed Anders and colleagues at the University of Chicago, with several other researchers later confirming its existence and panglobal extent. Even prior to the Chicago group's discovery there was other evidence in hand of extensive fires at the time, charcoal having been identified in some locations.

There is plentiful evidence available for an impact's having occurred at the time of the Cretaceous/Tertiary (K/T) extinction in which the dinosaurs disappeared, and for similar impacts at the epochs of other mass extinction events. What is especially pertinent here is the soot layer: How was it produced? It seems to indicate that the Earth went up in flames, or at least most of the plant life did. There are also paleontological data suggesting that although many marine creatures living in the top 10 meters of the oceans were wiped out, organisms living deeper down survived.

One suggestion for this has been that the dust thrown out from the impact site (or sites, since it is now widely believed that multiple impacts occurred at the time of the K/T extinctions) could have caused the ignition of fires through induced lightning. It is well known that lightning may be set off by volcanic ash falls, the settling ash causing charge separations to occur in the atmosphere, which are the source of lightning bolts.[1] With a huge impact, it is to be anticipated that there would be global fallout, and the iridium layer at the K/T boundary is evidence for just such an occurrence.

There are other reasons that one would expect the surface temperature of the Earth to be raised to an extremely high level, quite likely well above the ignition point of even a lush green tropical jungle, in the aftermath of a large impact. The Chicxulub Crater on the Yucatán peninsula of Mexico is at least 180 kilometers across, and there are some arguments that a more realistic diameter might be 400 kilometers. The problem is that 65 million years have elapsed since it was formed and such a huge crater on Earth quickly slumps under gravity so that a series of concentric rings is created. It is not always clear which of the rings was formed by a crater that was evacuated and then slumped and which by the outgoing shock wave. Dating of the crater indicates that it was formed at a time indistinguishable from that of the K/T boundary, the probability of that occurring by chance being much less than 1%. If this crater were produced by an incoming asteroid about 10 kilometers in size (as originally estimated for the K/T projectile on the basis of the amount of iridium found) meeting the Earth at about 25 km/sec, then the energy released is equivalent to about 100 million Mt. Such an impactor, however, would form a crater only 60 to 100 kilometers in diameter. Thus it seems that in the case of Chicxulub, the projectile was a lot larger—perhaps 20 to 40 kilometers in size—and the explosion much more energetic than originally visualized by Alvarez and colleagues.

The total volume of rock excavated from the crater was huge—at least 100,000 cubic kilometers and perhaps ten times that. Remember that the atmosphere of the Earth is very thin (in terms of thickness, as opposed to density). Atmospheric pressure drops to about one millionth that at the surface by the time you get to an altitude of 100 kilometers, a distance that is equivalent to half or less of the diameter of the crater. Thus the atmosphere is only a tenuous skin over the crater, and a large fraction of the pulverized rock ejected in the impact would be thrown up and out of the atmosphere on what is termed a *ballistic trajectory*. The rock is expelled upward and sideways, some of it moving swiftly enough to escape the Earth's gravitational pull, but most of it looping up above the atmosphere and reentering elsewhere around the globe. To take a conservative line, we will estimate that just 1% of the minimal 100,000 cubic kilometers of ejected rock is thrown above the atmosphere on ballistic trajectories, or about 1,000 cubic kilometers. Most will reenter in this way, the fraction escaping having implications for the concept of *panspermia*, the spreading of life from one planet to another. The sizes of the lumps of rock that will reenter range from that of sand grains to very large boulders, and most of these reentries would occur in the following few hours. It takes a weather satellite about 90 minutes to orbit the Earth at a low altitude, and the same laws of celestial mechanics apply to the ejected rock. For the Chicxulub impact, some rocks would have traveled just a few hundred kilometers from the impact site, landing in Haiti or Texas.[2] Some would have crossed the Atlantic to fall in the region where Europe now sits, while others would have performed a half or even full orbit around the planet. The point is that a wave of debris cascaded down upon the atmosphere from above, producing one of the greatest global meteor showers ever, although it is unlikely that the inhabitants below appreciated it. As the ejecta reentered, it would have burned up, producing an essentially unbroken wave of meteors above the flora and fauna below. The whole sky would have been filled with burning meteoroids, which an hour or so before had been rocks deep below the surface of the ground in a corner of Mexico.

How much energy will be irradiated downward by this rock as it burns up? At reentry speeds varying from a few kilometers per second up to 11 km/sec, 1,000 cubic kilometers of rock will release energy equivalent to about a week's worth of solar energy spread over the whole planet. In many ways, one can imagine the situation as being analagous to a huge griller being located 50 to 100 kilometers above the surface, boosting the surface temperature to over 1,000°C. It is only to be expected that under such circumstances the plant life of the continents would be rapidly desiccated and then ignited. No wonder there is a dense soot layer at the geological boundary at which the dinosaurs died. Computations by Owen Toon and colleagues at NASA–Ames Research Center show that the threshold for forest ignition globally would be reached with the heat produced by ballistic ejecta from an impact with an energy of about 100 million Mt, which is the minimum that seems to have been released at the time. An impact energy ten times lower would still produce ejecta capable of causing forest ignition over an area greater than that of the United States.

Much of the irradiation will come from the burn-up of smaller rocks and boulders, from millimeter to decameter sizes, at high altitude. We have already seen, however, that rocks as big as 50 to 100 meters in size will fragment and detonate in the atmosphere, as in the case of Tunguska. Can the Tunguska event tell us anything about the ignition of forest fires? Any visitor to ground zero in Siberia is amazed not only by the radial tree falls (the trunks lying so as to point away from the epicenter), but also by the fact that the trees are largely charred (but only on one side, that being the side directed inward). It is easy to imagine what happened. The 10- to 20-Mt explosion produced a sufficiently large radiative flux so as to ignite the trees even in the soggy swamps of the region—no magnifying glass needed here to concentrate the Sun's rays and get a fire going. Nevertheless, we know that the fire did not consume the forest in its entirety. Within minutes the blast wave reached the burning trees and blew the flames out, leaving them charred on one side, but not burned through.

There are several mechanisms, then, that might be expected to lead to global fires in the aftermath of a large impact. As explained earlier, at the K/T boundary the layer of soot has a thickness indicative of plant materials equivalent to at least 90% of the current worldwide biomass being incinerated. For small impacts, such as Tunguska, we have unarguable evidence of forest ignition, with other perturbations of the environment. For example, the evenings following the Tunguska explosion (on June 30, 1908) were noted as being "White Nights" throughout Europe, with people able to read newspapers from the sky glow even after midnight. To my knowledge, it has never been explained properly whether this was due to sunlight scattered by high-altitude dust, some form of induced airglow, or some unknown physical cause. There have been claims that dust from the shattered projectile is detectable in Antarctic ice layers that were laid down around then, which would imply global (not just hemispherical) spreading of the dust, but this has been contradicted by others. Such dust identification would be important because there is weak coupling across the equator between the circulations of the atmosphere in the northern and the southern hemispheres. If dust from Tunguska (which is near latitude 61° North) were found in Antarctica, it would demonstrate that even small impacts are capable of causing global perturbations of the environment.

The maintenance of a high-altitude dust layer would lead to the scattering away of some significant fraction of the incoming sunlight and thus a cooling of the climate. Again, there have been claims that the Tunguska explosion, minor though it was, caused cooling of the northern hemisphere by a degree or so in the few years following 1908, but because normal fluctuations induced by circulation patterns on the Earth, solar cycle variations, and so on, can cause deviations of the same magnitude, it is not possible to do more than point the finger of suspicion at Tunguska. The climatic effects of singular occurrences like Tunguska are not of great concern, since terrestrial phenomena such as volcanic eruptions can cause similar climatic excursions through the dust that they release into the atmosphere. It is of interest, however, to understand the climatic deviation caused by Tunguska for scaling up to larger impacts, these causing perturbations far beyond those producible by mundane cataclysms such as volcanoes.

In fact, it was recognition of the parallels between the smoke and dust produced by impacts like Tunguska and the similar atmospheric injections that would occur as a result of any nuclear war that led to the study of the climatic effects of such a war, and therefore the possibility of what has become known as a Nuclear Winter. As mentioned earlier, the energy released in the Tunguska explosion was, according to best estimates, equivalent to about 10 to 20 Mt of TNT. This compares to the most powerful nuclear device ever exploded by humankind, which was equivalent to about 67 Mt. Due to international treaties outlawing the atmospheric testing of such weapons, the natural

Tunguska explosion (and its effects on the environment) are of military interest.

The idea of nuclear winter is that, as a result of any nuclear exchange, there would be so much smoke injected into the atmosphere that an appreciable proportion of the incoming sunlight would be reflected away, cooling the climate. This is a very complex situation that is not as simple as just waiting for the smoke to clear and then expecting the climate to warm again to the original level; there are feedback effects that complicate matters. To highlight one feedback effect, as the climate cools due to the smoke shrouding the surface, a greater area will be covered with snow and ice. Snow is highly reflective, however, and with more snow covering the surface, less of the sunlight would be absorbed, thereby dropping the temperature still further and producing more snow cover, in a continuing cycle.

That is one scenario, but there might well be another equally likely, but diametrically opposite, effect, which we can consider by describing what scientists now think may have made Venus such an inhospitably hot place. Venus is thought to have become exceedingly hot due to a runaway greenhouse effect. The so-called greenhouse effect occurs to a certain degree naturally on Earth; otherwise, we would freeze to death.[3] In the case of the terrestrial greenhouse effect, the carbon dioxide, water vapor, and other constituent gases of the atmosphere prevent the Earth from cooling to the level that would occur if the planet were atmosphereless, which is some tens of degrees lower than the pleasant temperature to which we are accustomed. Since the temperature of the outermost layer of the Sun is about 6,000°C, it appears yellow-hot to us; that is, the peak of its emission spectrum is in the middle of that part of the electromagnetic spectrum to which our eyes are sensitive. Of course this is not by accident. Evolution has done its job. The solar radiation is partially absorbed by the planet, giving it a mean global temperature of about 10°C. At this temperature the Earth re-emits the absorbed solar energy at a wavelength that is in the infrared part of the spectrum and therefore undetectable by the human eye.

There is a simple relationship between the temperature of an object and the wavelength at which the peak occurs in its emission spectrum. That wavelength and the temperature are linked by the expression:

$$\lambda T = 2.9 \times 10^{-3} \text{ meter Kelvins}$$

where λ is the wavelength in meters and T is the temperature on the scale used by physicists (the absolute or Kelvin [K] scale). On that scale, the temperature is just the value in degrees Celsius plus 273, so that 100°C is equivalent to 373 Kelvins.[4] At a temperature of 10°C (or 283 K), the peak wavelength is just a little more than 10 microns, a micron being a millionth (or 10^{-6}) of a meter. At a temperature of 6,000°C, or 6,273 K, the peak wavelength is about 0.46 microns.[5]

In the infrared, the atmosphere is largely opaque, so that the re-emitted energy cannot escape freely. The infrared does trickle out, though, at certain specific wavelengths where there are gaps in the atmospheric opacity. This situation is similar to pouring water at a continuous rate into a bucket having several holes. The water level in the bucket is controlled by the number and size of the holes. If you plug one or two holes, the level will rise. So it is with the atmosphere, because the holes in the infrared transmission can be plugged by certain gases, and carbon dioxide (CO_2) in particular is important. Any rise in the CO_2 level (for example, due to our burning fossil fuels) will increase the opacity in some of the gaps and therefore limit the rate at which the infrared radiation can escape, causing the temperature to rise. This heating will produce a number of positive feedback effects. For example, at the higher temperature not as much CO_2 can dissolve in water, meaning that even more CO_2 is released from the oceans, where a substantial fraction of the terrestrial CO_2 currently is stored. In addition, more water vapor in the atmosphere will lead to other gaps being plugged, and more methane (released by such things as increased insect activity and faster turnover in marshes) might plug still others. Eventually a point can be reached where a runaway greenhouse effect occurs, and the temperature continues to climb under the positive feedback mechanisms. It appears that this has happened on Venus, without the aid of any biological activity, although not on Earth—at least, not yet.

Why are the Earth and Venus, often termed *twin planets* because of their similar sizes, so different? People tend to expect Venus to be hotter because it is closer to the Sun, but it is easy to show this to be a fallacy. Venus is cloaked with a perpetual bright cover of clouds, so it only absorbs about 20% of the solar energy radiated toward it. Venus orbits the Sun at a distance of about 0.71 AU. Because intensity drops off as the inverse square of the distance, the solar flux at Venus is about twice that at the Earth.[6] The two planets are within a few percent of being the same size, meaning that the total amount of sunlight hitting Venus is about twice that hitting the Earth, but the Earth absorbs more than three times as much of the incoming sunlight as does Venus. The latter is hotter not because it absorbs more solar energy, as one might think, but because it re-emits the energy less efficiently than does the Earth.

We now turn to what might have happened on Venus millions of years ago. The Earth and Venus are sister planets in terms of size and mass, and yet appear very different. Venus has a dense atmosphere (more than 95% CO_2), which, along with its thick cloud deck, has caused the runaway greenhouse effect and elevated temperature. People often ask why Venus has so much CO_2, but this is a question based on a false premise: The Earth has just as much CO_2, but it happens that our CO_2 is locked up partially in solution in the oceans and in plant life but mainly as carbonate rocks, which require water for them to be formed. The question that therefore needs asking is, "Why has Venus little or no water?"

We know that in the first half billion years after the planets formed, the number of gargantuan asteroids (or comets, or planetesimals; call them what you will) was very high. The result was that each of the planets and the Earth in particular was impacted often. It is easy to see that this must have been the case. Looking at the lunar surface through a telescope one can see five large "seas," or *maria*, which are huge basins excavated by major impacts early in the Moon's history and later filled with lava. Such impacts were close to being energetic enough to split the Moon asunder. Scaling up to the Earth, taking into account our larger mass and radius, one expects at least 50 such impacts to have occurred. Each of these had sufficient energy to evaporate all the water currently seen on Earth and likely a substantial fraction of the crustal rock as well. Once the water was vaporized, it would form a massive atmosphere that would be subject to gradual loss into space, because the water molecules would be dissociated steadily into their constituent hydrogen and oxygen atoms by the solar radiation of ultraviolet light, escape of the light atoms into space then being possible. Indeed, there is a continual loss of atoms from the top of the Earth's atmosphere. Moving up through the atmosphere, the different levels are given different names, such as the troposphere, stratosphere, mesosphere, ionosphere, and thermosphere. The topmost zone, from which atoms can escape to space, is called the exosphere; that is, the region from which particles are making their exodus.

Although some later water liberation from the interior of the planet, for example through volcanic action, could have contributed some significant fraction, it is thought likely that the majority of our water was supplied through smaller cometary arrivals—comets broadly similar to those that we observe now—after the major impacts had ceased, starting some time around 500 million years after the Earth aggregated but continuing to the present. In other words, although very large impacts may result in an overall mass loss from a planet, through rocks being ejected and water evaporated and then lost through the exosphere, smaller impacts can lead to an accumulation of matter. The Earth seems to have built up a watery environment gradually (the hydrosphere) after a tumultuous early history in which the water complement was often knocked back by very large impacts. Since around 3.8 billion years ago, that water has been responsible not only for making life possible, and thus providing a positive feedback mechanism due to the plant life removing CO_2 from the atmosphere, but also for dissolving CO_2 and laying it down as carbonates, thus limiting the greenhouse effect. If we did not have water, we would be much hotter than we are, although paradoxically water vapor also acts as a greenhouse gas, plugging some of those holes in the infrared opacity.

But what about Venus? It seems that on Venus things never came under control—that enough water never settled to start the process of limiting the CO_2 in the atmosphere. There is no reason to suspect that Venus's early history (gigantic impacts scouring it of water) would have been significantly dissimilar to that of the Earth. There would be a few differences (for example, due to the mean impact speeds on Venus being a little higher than on the Earth), but it seems unlikely that this could explain the radical differences in their evolution. However, the later impacts by smaller comets, through which water accumulated on Earth, could have been less efficient in supplying water to Venus. This might have been due to the fact the impactors hit Venus at higher average speeds and might therefore have caused atmospheric erosion rather than accumulation. Impacts were probably also less frequent on Venus, which might have been a contributing factor. And, the comets that hit Venus generally orbited the Sun at closer range; these comets lose their water more quickly than those that do not approach the Sun so closely, meaning that the amount of water present in the average comet hitting Venus would be less than in the case of a comet of the same size hitting the Earth.

Although all three of these factors were probably of minor effect in their own right, it could well be that, taken as a whole, they explain the present state of Venus. It is not so much a case of Venus's being a planet that went wrong, as the Earth being a planet that came right, at least from our perspective. Both would have been stripped of their water in the early history of the solar system through giant impacts compounded by the extremely hot environment immediately after the planets formed, and through later impacts by smaller water-rich comets being responsible for the build-up of the terrestrial hydrosphere and, one way or another, the absorption of the carbon dioxide. It seems that Venus did not quite make it, because it received water at a rate insufficient to lead to its accumulation and consequent counteraction of the greenhouse effect. Alternatively, perhaps the huge impacts on Venus in its early history led to the conditions never being right for water to later accumulate, even though the influx might have been the same as to the Earth. There is some water on Venus, but not much. The Soviet Venera probes detected traces in the Venusian clouds as they descended to the surface, as did the several atmospheric probes in NASA's Pioneer Venus mission in 1978. The point here is that any planet may be positioned on a knife-edge, feedback effects being capable of causing a runaway in either direction: to low temperatures, as in the case of a nuclear or cosmic winter when dust shrouds the planet, or to high temperatures, as in the case of a runaway greenhouse effect.

Having brought up the idea of the greenhouse effect on the planets, and also having discussed global fires following a large impact, one might put two and two together: Although the smoke released can cause planetary cooling because it scatters sunlight away, equally well the carbon dioxide emitted by the burning forests will boost the amount of that gas in the atmosphere and induce an enhanced greenhouse effect. It is also to be anticipated that after a large impact there will be CO_2 release from carbonate rocks, accentuating any post-impact greenhouse effect. Environmental modeling of the aftermath of a giant impact indicates that the greenhouse effect induced by CO_2 release may persist for 10,000 years or more, this being the longest-lived stress on the environment (see Table 3). Interestingly, the Chicxulub impact linked to the K/T boundary occurred on carbonate rock, suggesting that a severe greenhouse phase might be expected. This is borne out by studies of the climatic perturbations.

One might expect the smoke to act in a similar way to dust suspended in the atmosphere, reflecting away sunlight and causing cooling, but just to confuse the matter it has been found that the smoke from the fires may actually cause overall heating. It all depends on the height above the ground that the smoke reaches. High-altitude smoke will tend to absorb sunlight (the smoke particles having very low albedos), producing heating of the air far above the surface. That heat is easily radiated away as infrared because there is no thick atmosphere above to absorb the emissions. This would lead to a global cooling. However, lower altitude smoke will absorb the solar radiation in the same way, causing cooling at the surface (which no longer receives the sunlight directly) but overall global heating due to a larger fraction of

Table 3
Environmental Stresses Produced by Large Impacts, Such As That at the K/T Boundary

Stress Induced	*Time Scale*
Darkness (loss of photosynthesis)	Months
Cold (impact winter)	Months/years
Winds (500 km/hr plus)	Hours
Tsunamis	Hours/days
Greenhouse (H_2O)	Months
Greenhouse (CO_2)	10,000–500,000 yrs
Fires	Months
Pyrotoxins (poisons produced by fire)	Years
Acid rain	Years
Ozone layer destruction	Decades
Volcanism triggered by impacts	Millennia
Mutagens	Millennia

the incoming sunlight being absorbed (that is, the global albedo is lowered).

If you are confused by all of this and cannot see whether to expect an impact to cause a heating or a cooling, don't be worried. The scientists involved are also in dispute as to exactly what to expect to happen. The majority view is that for the first few days or weeks there would be a gross heat pulse afflicting the Earth, due to the reentering ejecta and then global fires. After that, things would get much cooler, the pulverized rock and dust left in the atmosphere shrouding the planet and raising its albedo so that less sunlight is absorbed. This would last for some months or years: It takes a long time for submicron dust to settle out. After the settling, the elevated carbon dioxide levels would take over, causing a severe greenhouse heating that would persist for many thousands of years until the Earth returned to something like its pre-impact state. Of course, in the meantime the biological balance of the Earth would have been grossly upset. The important point is that one way or another, the global climate will have been severely disrupted, leading to radical changes in the populations of flora and fauna on land and in the oceans.

Much has been made of the comparison of a global nuclear war to the catastrophic impact scenario—especially regarding the onset of a nuclear winter—so it is instructive to consider now the accuracy of this comparison. In doing so, consider again the nuclear winter scenario. Toon and colleagues found that their moderate nuclear winter model would lead to sufficient smoke being lofted such that the light level at the surface would drop by about a factor of 10, leaving the brightness at a level similar to that on a very cloudy day. To get a better sense of how much light this is exactly, consider the following comparisons. Relative to the sunlight received on a clear day (valued as 1), the following are the levels of light at which different phenomena occur:

Very cloudy day	0.1
Limit for photosynthesis	0.01
Full Moon brightness	0.000001
Limit for vision	0.0000000001

To offer further comparison, relative to the predicted drop in light in a nuclear winter scenario, recent major events, such as the Mount

Pinatubo volcano in the Philippines in the early 1990s and the oil fires in Kuwait as a result of the Gulf War, have had a very restricted effect, dropping the mean sunlight brightness at the surface by just a few percent and only for a limited time. This is not to say, however, that these had no effect on the climate: Drops in global temperatures have been claimed at about 0.5°C for 2 years from Mount Pinatubo, with a similar effect from the El Chichon eruption. The nuclear winter drop in light level by a factor of 10 calculated by Toon and colleagues is much more severe, however, being sufficient to provoke a mean temperature drop bad enough to cause loss of a complete growing season. There would also be a negative feedback effect: The snow and ice formation discussed earlier will set in such that the winter will continue even after the smoke has dispersed. That is, an induced ice age will result. That's for the case of a nuclear war; now, what about the impacts?

If the threshold for an induced winter (nuclear or impact) is a drop in light intensity by a factor of 10, then it is quite straightforward to calculate the amount of dust needed to give the same effect as that already described. The answer is about 10^{10} tons in all, which is equivalent to a body about 1.5 kilometers across. What we need to know is how big the explosion must be in order to loft this mass of submicron dust into the stratosphere. Measurements of nuclear explosions have shown that each megaton of energy elevates about 200,000 to 500,000 tons of dust to the stratosphere, and of that about 10% is submicron, so it remains there for some months at least. To get 10^{10} tons of submicron dust into the stratosphere therefore requires, given the figures provided here, an explosion with an energy yield of about half a million megatons (in round figures). This implies an impactor of somewhere in the 1- to 2-kilometer region; that is, an impact winter would result from the collision on the Earth of an object around 1 to 2 kilometers in size. There can be little doubt, therefore, that an impact by such a projectile will lead to global effects of the type I have described.

The maximal threshold size that can cause a global catastrophe has been discussed; *maximal*, in that any other environmental perturbations that are caused will add to the disturbance of the climate and therefore lower the threshold to perhaps somewhere below 1 kilometer. For example, the aforementioned calculation was based on the dust only, whereas we noted that smoke, gram for gram, has a much greater effect on the attenuation of the sunlight reaching the ground. A 1- to

2-kilometer impactor will throw out ballistic ejecta capable of igniting forest fires over an area up to a million square kilometers. If the impact were in the Sahara desert or the center of Australia, this might not be a great problem with regard to the smoke produced, because the biomass per square kilometer is low; but if the impact were in the Amazon basin, much of the dense rainforest would be ignited, leading to a smoke-loading of the atmosphere far in excess of that imaginable from man's burning of the rainforests or the oil wells of Kuwait.

There are a number of other potentially catastrophic atmospheric effects that might be caused by an impact that have not yet been mentioned, apart from being listed in Table 3. In addition to the post–industrial revolution greenhouse effect, one of the other main environmental concerns of the present day is the destruction of the ozone layer by manufactured pollutants. It has been estimated that the Tunguska explosion caused the destruction of 30% to 50% of the ozone over the northern hemisphere; such effects are of direct concern because this ozone regenerated in a matter of days, whereas chlorofluorocarbons and other nonnatural pollutants act as catalysts that continue to break down ozone, leading to a long-term diminishment in its concentration. Therefore, we should ask whether large impacts can have atmospheric effects beyond the straightforward cooling by submicron stratospheric dust considered earlier.

Our atmosphere is composed predominantly of nitrogen and oxygen molecules (about 78% of the former and 20% of the latter), but with very few nitrogen oxides (usually written as NO_x because various combinations can occur). The small amount of NO_x can have various noxious[7] effects and play a major role in disturbing the atmospheric chemistry—for example, in dispersing the ozone layer at tens of kilometers' altitude—if it is deposited high enough. A 100,000-Mt impact produces in its plume sufficient NO_x to cause global removal of the ozone screen, as long as the NO_x is concentrated in the stratosphere and mesosphere. If the NO_x were mixed throughout the atmosphere, a larger impact, about a million Mt, is required. Under either circumstance, the solar ultraviolet flux reaches the surface largely unimpeded, with obvious consequences for the lifeforms below (although I have yet to hear anyone suggest that the dinosaurs died of skin cancer). The NO_x is not quickly removed from the upper atmosphere, leading to a sustained period of ozone depletion. As it is dispersed and eventually rained out, it has other deleterious effects on the environment; for example, NO_x causes acidification of the oceans, the hydrosphere being deluged with dilute nitric/nitrous acid. There is a parallel here with present-day concerns over acid rain, which is predominantly produced now by the sulfur dioxide/trioxide released into the atmosphere as we burn fossil fuels. In an impact, SO_x released from sulfates in the target rock has been identified as another major agent causing atmospheric/hydrospheric acidification.

Note that we have been considering only impacts close to the threshold at which a global catastrophe would occur. The catastrophe in question here—an impact-induced winter that will lead to the cessation of plant growth—is moderate compared with the consequences of much bigger impacts (100 million Mt and worse) causing mass extinctions. With a more moderate impact, plant growth may halt for a year, but the seeds will survive; we are not talking about an extinction event. Earlier in this chapter, we looked at the global wildfires that apparently followed the huge impact(s) at the K/T boundary, raising the temperature globally for at least some days or weeks. This was followed, however, by a marked cooling according to climatic models of the effects of the resultant dust cloud, with continental temperatures falling by up to 40°C for a year or so, the oceans by rather less since they are buffered by their large heat capacity. If smoke from the burning of the global biomass is included in this modeling, the effects are more severe. What happens in the case of a 1- to 2-kilometer asteroid impact?

The global cooling leads to a cessation in the normal hydrological cycle because the rate of ocean evaporation plummets. This results in a global drought lasting many months as rainfall is reduced to less than 10% of its normal value. More sophisticated modeling—which does not necessarily mean "better," since there is still much that we do not understand about the effects of severe climatic perturbations—indicates that an impact by a 1-kilometer asteroid would cause a dip in global land temperatures by about 8°C by about 15 days post-impact, with a gradual rise back to the norm within a couple of months. A 2-kilometer impactor would cause a 10°C drop and a more gradual rise back to normal, while a 10-kilometer asteroid would result in a 15°C drop, which is sustained for much longer. This appears to confirm the earlier calculation, which was based solely on the mass of dust required, that the threshold for a severe climatic disruption resulting in mass starvation is somewhere in the 1- to 2-kilometer region.

Much has been made here of the role of dust deposited high in the atmosphere in causing a global cooling of the Earth in the aftermath of a huge impact. For the dust to be injected at a sufficient altitude to have this effect, such that it circulates about the planet and does not quickly fall out, the impact must be extremely energetic. It is also possible for dust mantling of the atmosphere to occur directly from space, however. We have stated that the total amount of submicron dust needed is about equivalent to the total mass of a 1.5-kilometer asteroid or comet. Dust of such size does not burn up in the atmosphere, because grains smaller than about 50 microns are decelerated without melting or ablating. Is it possible that dust veiling of the Earth directly from space could cause global cooling without an actual impact's occurring?

The first thing to note is that an object much larger than 1.5 kilometers would be required, because only a small fraction of the dust it liberates would make its way to the Earth. However, one possible scenario would have a comet larger than 10 kilometers or so being injected into an orbit in the inner solar system, breaking up, and producing a large quantity of dust that remains in heliocentric orbit.

In fact, the Earth does move through a cloud of dust called the *zodiacal dust*. The cloud gets its name because it gives rise to the zodiacal light, a diffuse glow in a huge triangular shape which follows the path of the Sun across the sky.[8] This phenomenon is best seen an hour or two after sunset or before dawn, from a location in near-tropical latitudes. Few people have consciously observed the zodiacal light because it is necessary to be well away from city lights in order to see it. That is not to say that its overall light power is small, however. In fact, 40% of the integrated light in the night sky is from the zodiacal dust cloud. The origin of this diffuse glow is the scattering of sunlight by dust grains in space, which mostly range in size from 10 to 100 microns. Until recent years, it was only the zodiacal light that gave us a reasonable indication of the distribution of small particles in the inner solar system, apart from a few spacecraft experiments which had rendered some data by counting impacts as they followed paths through interplanetary space.

One of the surprises from the European Space Agency's *Giotto* space probe, which flew past Comet Halley in 1986,[9] was that there was a much larger number of very small dust grains than had previously been anticipated. Attached to the shield covering all of one side of the satellite, protecting sensitive instruments below from meteoroid impacts, Tony McDonnell and his team at the University of Kent in England had attached a series of detectors similar to small microphones. These were designed to register the pings as dust grains smashed into the shield while *Giotto* hurtled by P/Halley at over 70 km/sec. In fact, during the closest approach phase, during which the probe got to within 600 kilometers of the comet nucleus, flying deep through the cometary coma, McDonnell's detectors were counting impacts at a rate of over 100,000 per second, most of them by submicron grains.

It seems, then, that comets produce more submicron grains than expected. These would not show up in zodiacal light observations because they scatter too little sunlight to be discernable. However, returned exterior surfaces from satellites in orbit about the Earth (such as NASA's Long Duration Exposure Facility [LDEF], which was plucked from orbit in 1990 after spending nearly 6 years in space) tend to confirm that the population of very small particles is much higher than earlier models had predicted. These may originate directly from comets, but also through the shattering of larger zodiacal dust grains as they happen to hit each other at hypervelocities in space.

The coma of a comet may be 100,000 kilometers or more across, and its dust tail millions of kilometers long. Because that coma size is about ten times the diameter of the Earth, one would expect close passages of comets by the Earth, resulting in a transit of our planet through the coma (but not an impact by the nucleus) to occur perhaps a hundred times as frequently as cometary impacts themselves. Transits through the dust tail will be even more frequent because the tails present even larger targets. The possibility therefore exists of dust mantling of the Earth directly from cometary dust soon after its release. Such events will occur much more frequently than big impacts. The amount of dust swept up in each transit through a tail/coma will generally be too small to produce catastrophic climate deviations, but they could cause temperature fluctuations of a few degrees, and have important effects on agriculture in marginally cultivatable regions. In the case of a massive comet, a near-miss of the Earth could cause a catastrophe. In 1978 British astrophysicists Fred Hoyle and Chandra Wickramasinghe suggested that the K/T extinctions were due to such an event. They have since reiterated the view that more modest episodes could be having a contemporary effect that should not be ignored in greenhouse modeling.

We do, in fact, effectively pass through comet dust tails many times each year. This is when meteor showers occur. These occur systematically, for example, in the second week of August when the Perseid shower is seen in the northern hemisphere. In such cases, the larger particles (meteoroids) released by the comet have become dispersed more or less equably about its orbit, forming a stream. And so, a shower is seen each year as our planet passes through that stream, rather than only in the years when the comet passes through the inner solar system. Other examples are the showers known as the Eta Aquarids (first week of May) and the Orionids (third week of October), which are known to have originated from P/Halley. You therefore have a chance to see pieces of Comet Halley twice each year, rather than having to wait until 2062 when the comet will next return.

If we recall the start of this discussion, where we were interested in the effect of dust accumulating in the atmosphere of the Earth directly from a large comet, we can see that there is a danger to our climate posed by any comet arriving in an orbit with perihelion in the vicinity of, or just beyond, the Earth. If this comet attains a stable subjovian orbit, it will be particularly dangerous because it will remain in that orbit for many millennia, releasing dust and meteoroids. The dust particles soon will have their orbits shrunk, producing a thick cloud through which the Earth will pass continuously, and the particles smaller than about 100 microns will accumulate intact in the atmosphere. Of these, the larger grains soon settle out, but the smaller ones remain aloft for months or years, veiling the atmosphere and leading to global cooling.[10] The meteoroids will form a stream that will impinge on the Earth annually (or biannually), causing climatic perturbations through upsets of the atmospheric chemistry. The decay of this extensive meteoroid stream and dust cloud takes up to about 100,000 years.

This may sound like a wild story predicated upon an arbitrarily invoked event: a very large comet trapped in an orbit within that of Jupiter, but with perihelion somewhere near the Earth. But it is a scenario for which there is hard evidence. As described in the next chapter, although the iridium layer at the K/T boundary is often described as being so thin that it must have been deposited at essentially an instant in time, this is only a *geological* instant. It is possible that the iridium may have been laid down over a quite extensive period (many years). Some geologists invoke this possibility to imply that the iridium may have been released by volcanic action. The reason that the iridium layer at the K/T boundary was initially ascribed to a large asteroid impact is that this metal is rare in the terrestrial crust but common in meteoritic material. Iridium is, however, more common in deeper layers of the Earth and so might perhaps be liberated at the surface in major volcanic events.

But there is other evidence that indicates that the iridium had an exogenous source. Apart from the thin iridium deposition, around the K/T boundary there is a layer, at least as thick as would be deposited over 50,000 years or more, of amino acids known to have an extraterrestrial origin. These would have been destroyed in any large impact by being split into their individual atoms; that is, if they had arrived in an asteroid or comet, they would have been lost from the record. Thus it appears that they must have arrived in the form of dust that was gently decelerated in the upper atmosphere, gradually settling to the surface and being incorporated into the strata. This dust accumulation continues up to a period of about 100,000 years. This period happens to coincide with how long it would take a complex of asteroids or comets, meteoroids, and dust produced by a giant comet break-up to be cleared from the inner solar system. Coupled with the fact that there are at least two other craters known with a similar antiquity to Chicxulub (that is, it seems that there was more than one impact at the time of the K/T events), this indicates that the original concept of the Alvarez team—a behemoth of an asteroid coming from out of the blue—is incorrect. Certainly, there were one or more immensely powerful impacts—perhaps there are other Chicxulubs awaiting discovery—but there was a lot more happening in the sky at the same time.

The important point to note from this discussion of what happened 65 million years ago—and what may be happening now—is that reality is just not as simple as the idea that there are a number of big objects in orbits that cross that of the Earth, and every so often one of them happens to blast into our planet in a calamitous event. The ways of the universe are much more subtle than that. The daughter products that they liberate, from dust grains to smaller asteroids, can also affect the environment, and hence life on the Earth.

5

Holes in the Ground: The Evidence for Past Impacts

Now that we know the possible effects of an impact, we turn again to the questions of whether such impacts really happen here on Earth and, if so, what evidence there is for them. Returning to our view of the Moon through a telescope, clearly there are myriad craters spread over its surface. Many bizarre suggestions were made for their origin over the decades leading up to the 1970s. Even given that their impact genesis was confirmed through the analysis of lunar samples returned in the Apollo and Luna programs,[1] there could be grounds for complacency in regard to the danger to humankind, if one is fooled into believing that the craters are due to collisions in far antiquity. In fact, although it is known that many of the largest craters were created within the first billion years of the Moon's history, there is ample evidence that craters of all ages exist, with some having been produced very recently. How recent is *recent* is a moot point, but certainly the seismometers left on the lunar surface by the Apollo astronauts registered many impacts by meteroroids with masses of a ton or more during the 1970s. Larger impacts on the Moon have also occurred within the past few million years. This can be judged by the existence of craters with bright ejecta rays; these have not had time to be darkened by the billions of impinging meteroroids and dust which churn over the uppermost layers of the lunar soil. This action is termed *lunar gardening*, for obvious reasons. These bright ejecta blankets also have a sparsity of superimposed small craters, compared with the surrounding older terrain,[2] proving their youth.

It has been mentioned that the Moon, and indeed all of the terrestrial planets, were subjected to a high impact rate early in their histories. This phase is termed the *late heavy bombardment*, in which many collisions occurred with the large number of planetesimals (by which term one could understand *comets* and *asteroids*) left over from the formative stages of the solar system, in which millions of such bodies agglomerated to form each of the planets. Indeed, such a bombardment could be thought of as the final stages of the accretion of the planets, continuing through until about 3.5 to 3.8 billion years ago. The word *late* here implies later than the time when the planets formed, 4.5 billion years ago. The lunar maria—the large dark plains on the Moon that were once thought to be seas before we understood the Moon to be waterless—are the huge basins formed as the result of impacts by very large planetesimals, later filling with basalt from volcanic flows. Such volcanism terminated on the Moon billions of years ago, after it cooled from its original high temperature, a temperature that was partially attained from the huge quantities of energy released in the impacts of planetesimals forming the maria. Scaling from the size and mass of the Moon, it is to be expected that about 50 impacts occurred on the Earth by planetesimals of comparable dimension to those that produced the basins of the maria. Each of these giant impacts would have been sufficiently energetic not only to cause the loss of all surface water on the Earth, but also to vaporize part of the crust. The start of continuing life on Earth would have been frustrated until after the cessation of this late heavy bombardment, since such huge impacts would effectively sterilize the planet. Thus life may have originated many times before the conditions were right for its proliferation and continuation.

With a mentality determined to deny your own mortality, and even the mortality of the human race, it might have been possible before about 60 years ago to explain away the chance of a gigantic impact causing a catastrophe on the contemporary Earth. Sir Edmond Halley had spoken about the consequences of a comet hitting the planet, but because almost all comets known to exist before the start of the twentieth century had orbits that crossed Jupiter,[3] they could be explained away as being much more likely to hit that planet, saving us from catastrophe: The likelihood of a typical comet's striking Jupiter is about a thousand times as high as the Earth-impact probability. In addition, studies of the orbital evolution of such bodies had shown that Jupiter (in particular) could remove comets by ejecting them from the solar system rather than just swallowing them up. In the late nineteenth century, the American astronomer Hubert Newton derived analytical formulas that showed that the chance of a close passage past Jupiter that causes a comet to be ejected is about a thousand times as high as the chance of an impact occurring.[4] This implies that comets in general do not remain on Earth-crossing orbits for long, timescales of the order of 10,000 years being appropriate before ejection might be expected. If the probability of a particular comet hitting the Earth were only about one in a billion per orbit, there would be a very small chance of that comet hitting the Earth before it passed close enough by Jupiter to be thrown out of the solar system, never to return again. Of course, this line of reasoning is fallacious because a steady number of comets on Earth-crossing orbits must lead to impacts from time to time, even if only a small fraction of them end their lives that way, but it is easy to convince oneself when so much is at stake. No one really wants to believe Chicken Little's warning that the sky is falling.

This fallacious belief in the security of the Earth gained support from the remarkable circumstances of Comet Lexell, which was discovered in 1770. This comet was first observed by the French astronomer Charles Messier[5] on June 14, and a little over two weeks later it made the closest-ever observed approach by a comet to the Earth, missing the planet by 0.015 AU (2.25 million kilometers, or about six times the lunar distance). This comet takes the name not of the discoverer Messier (thus contravening modern naming rules), but of St. Petersburg astronomer Anders Johan Lexell (1740–1784), who showed that the comet was not of long period, but quite the opposite: It had the very short orbital period of 5.6 years. Further, Lexell considered why the comet was not discovered before 1770 and calculated that an approach to within 0.1 AU of Jupiter in 1767 had caused a major change in its orbit about the Sun.

Prior to that diversion, the comet was too far from the Sun to have been seen on an earlier orbit. Having considered what had happened previously, Lexell then turned his attention to the future orbit of the comet. The important point was that its orbital period of 5.6 years was almost exactly half that of Jupiter's, implying that although the comet was safe for its next orbit (although it was not observed in 1776 due to the relative geometry of Earth, Sun, and comet being uncooperative), in 1779 another close approach to Jupiter was likely. The reason was that both the comet and Jupiter were returning to the positions they had occupied in 1767. In the event this was indeed what happened, and the comet was thrown onto a long-period orbit—it may be back in the twenty-second century. Humankind, then, had received two contradictory messages: (1) that comets could pass near the Earth, as had Comet Lexell in 1770, meaning that impacts were possible, and (2) that the Earth/humankind could be saved from impacts through the agency of Jupiter's regularly sweeping the inner solar system clean of such hazards.

Such was the general view during the first centuries of modern science regarding how extraterrestrial factors influenced the terrestrial environment. The perception was that they do not affect the Earth at all. This is quite the opposite of the ancient view, which in most cultures perceived terrestrial events to be closely connected to celestial portents. Essentially, the prevailing view for much of the past few centuries has been one that is termed *uniformitarian*, whereby catastrophes that could be postulated as originating through the effects of impacts by comets—no Earth-crossing asteroids yet having been discovered—were argued simply not to occur. More precisely, they were not argued about at all, but were generally ignored. The opposing view, that of *catastrophism*, was rejected on the basis of an interpretation of the geological and biological record, which held that all observed phenomena (such as the changes between geological epochs and the alterations in species from one era to the next) could be explained in terms of gradual changes rather than the abrupt modifications that would occur under impact-driven environmental stresses.

This gradualist-catastrophist schism has wider repercussions than merely the effects of powerful impacts. Even a cursory glance around can show that it is infrequent catastrophic events rather than the long-term action of gradual changes that largely control the face of the Earth.

For example, I recall once driving down from Tucson to Tombstone, Arizona, to see the famous OK Corral and Boot Hill. As I watched the baked scenery around me, I marveled at the deep channels scoured out of the arid desert, wondering how they could have been produced. On the drive back, I found out: An intense thunderstorm resulted in those channels being flooded by a torrent of water, scouring them ever deeper. By the time I reached Tucson, the storm was all over, with barely a rainbow or a puddle to mark its passing. The point here is that the conditions observed for 99% and more of the passage of time do not inform you about what is responsible for producing the geomorphology that you witness. Equally well, those storms are of phenomenal import for the evolution of the life forms found in the desert, showing that rare inundations may be shaping the biological, as well as the geological, scenery. Other types of cataclysm undoubtedly also play a part.

Historically speaking, although there were other lines of terrestrial evidence for catastrophism, it was often the extraterrestrial evidence from astronomical observations that would lead to a resurgence in popularity of such ideas; for example, due to the several spectacular meteor showers or storms seen during the nineteenth century. This is not to say that there were no celestial events that did not benefit the uniformitarian school; for example, the disintegration of Comet Biela in the 1840s led to the spectacular Andromedid meteor storms in 1872 and 1885, but no terrestrial catastrophe, despite the doomsayers. Similarly, the widespread panic concerning our passage through Comet Halley's tail in 1910 was soon seen to be groundless. It should not be imagined that in the late twentieth century our more sophisticated culture has ceased responding to singular events in a breathless way. The issue of impacts is now on the table due to four main causes: (1) the widespread acceptance that Meteor Crater and a few other well-known craters have a similar origin to the lunar craters, (2) the recognition that the Tunguska event in 1908 was indeed an example of a cosmic impact, (3) the recognition that the dinosaurs' demise 65 million years ago can be linked to one or more impacts, and (4) the discovery from time to time of objects that have had near-misses of our planet. The impacts of the fragments of a comet into Jupiter in July 1994 have given the debate another strong push. In the nineteenth century, however, the publication of Charles Darwin's *Origin of the Species* in 1859 (in which evolution was described as a progression of gradual changes) and the subsequent famous debate between Thomas Huxley and William Wilberforce (the Bishop of Oxford) at the British Association for the Advancement of Science in 1860 led to the nails being hammered into the coffin of catastrophism for more than a century. As we have seen, many of them are yet to be prised out, their burial being made more secure by the "cataclysms that weren't," in 1885 and 1910, for example.

Pulling such nails is much like pulling teeth from the heads of the supporters of the uniformitarian view. Although the public may view science as being a pursuit by high-minded individuals who are directed toward revealing the truth of how nature works, in fact it is often wracked with acrimonious and personal controversies, with the powers-that-be at any time behaving as effectively as the "Thought Police" of fiction. It is often possible for a particularly powerful person or clique to continue to control an area of science against all reasonable argument, not only until they retire, but even unto death and beyond. Thus the discovery of the first Earth-crossing asteroids in the 1930s did not herald an immediate revolution in traditional thought, although the arguments outlined earlier could no longer be invoked. These asteroids had orbits well within that of Jupiter, implying that their removal must be by impact in almost all cases, with the Earth and Venus being more or less equally likely to be hit. Further, their short orbital periods meant that their terrestrial impact probabilities are high on an annual basis (unlike comets that pass perihelion more infrequently), and they had orbital planes close to that of the Earth and were dark and fast-moving, making them difficult to detect and implying that there might be many more awaiting discovery—all making contributions to any estimate of the Earth-impact rate. This should have jolted scientists into attention, the origin of the lunar craters then being obvious, but the few who did point this out, such as Watson and Baldwin, were largely "whistling into the wind."

Science often moves slowly, especially when the accepted ideas are deeply entrenched, as was the case in the uniformitarianism versus catastrophism schism. Even by the 1960s, the idea that impact craters were a common feature of the Earth's surface, and an important geologic process, had hardly permeated the geological community, let alone gained widespread acceptance. This was the situation as the space age began and the Apollo program got under way. To which extent, and with the benefit of hindsight, one can say that the major factor shaping the face of the Moon—impacts, of course—was not, at that stage, part of the psyche of geologists in general. They were therefore unprepared to ask the correct questions of the samples later returned during the Apollo program. This situation was exacerbated by the fact that the Apollo program coincided with the geological community's trying to take on board the idea of plate tectonics, or continental drift. This debate, which is not yet over in a few quarters, had been simmering ever since the concept had been suggested several decades earlier. Although many a school child has looked in an atlas and noticed that the northeast of South America fits snugly against the concavity of Africa, opposite in position across the expanse of the Atlantic Ocean, geologists were askance, even apoplectic, at the suggestion that at one time these land masses may have been in contact. Ideas such as Baja California "surging" north and being in some way responsible for the earthquakes along the San Andreas and other faults in California, or the Indian subcontinental plate "ploughing" into Asia and causing the buckling-up of the Himalayas, were still far from being palatable, let alone swallowable.

Leaving plate tectonics aside, we return to the history of the hypothesis that Meteor Crater has an impact origin to see how the acceptance of large impacts on the Earth has gradually come about. There were earlier suggestions that specific geological features on Earth might have been formed by impacts. Sir Edmond Halley had mused upon whether the Caspian Sea and similar vast lakes might have been formed by cometary impacts. In the late nineteenth century, one physicist who thought that impacts could be important processes helping to shape the face of the Earth was Alexander Bickerton, Professor of Physics and Chemistry at the Christchurch campus of the University of New Zealand. One of his students was Ernest (later Lord) Rutherford, who will be best known to most readers as the man who split the atom. Bickerton suggested that, among other topographical features, the curved form from Cape Horn to the Antarctic Peninsula through South Georgia, the South Sandwich Islands, the South Orkney Islands, and the South Shetland Islands was due to a huge impact at a glancing angle.

For Meteor Crater itself, Daniel Barringer is probably remembered as the most ardent proponent of the theory that it was formed by an impact. Barringer was a mining engineer who actually bought the crater because he became convinced that if the crater were mined, a huge lode of iron and nickel, left by the impacting asteroid, would be found there. He spent huge sums of money drilling into the crater floor and walls to no avail, but he remained certain that the site was an impact crater. He pressed through with this suggestion in the face of rock-hard resistance from the United States Geological Survey (USGS). It was not Barringer, however, who first considered the possibility that Meteor Crater was created in this way. The first person to consider an impact origin for the crater was none other than the chief geologist of the USGS, Grove Karl Gilbert, who adopted the idea as a working hypothesis in 1891 when he went to Arizona to inspect the structure. He was subsequently influenced to reject this hypothesis.

It is important here for the reader without a background in scientific research to understand the nature of scientific proof, which is contrary to what might be literally termed *common sense*. The ways of science are very different from those of the judiciary. A working hypothesis is proposed, and the scientist then tries to disprove its correctness; that is, there is no *proof* in science, only *disproof*. In this simple situation, Gilbert's working hypothesis was, "Meteor Crater has an impact origin," which he then answered in the negative (but for erroneous reasons, as will be seen). His assumptions were wrong, not his hypothesis. This is the way in which science is supposed to progress, but a lack of adherence to this simple principle often leads people astray. A brilliant example of the application of this working hypothesis concept was given by Luis Alvarez. Alvarez investigated the physical evidence (that is, those parts of the evidence that are amenable to investigation through the application of the laws of physics) of the assassination of President Kennedy, driven largely by oft-repeated—and erroneous—statements along the lines, "The backwards motion of JFK's head as the fatal bullet struck proves that at least one shot must have come from the front, since a shot from behind would be against the laws of physics." Alvarez adopted the hypothesis, "The laws of physics require the shot to have been from the front," and showed this hypothesis to be false; that is, the shot could have been from behind. This does not *prove* that the shot was from behind, but it does *disprove* the presumption (frequently made by those who know little physics) that the shot must have been from the front. Confusing though it is to the ingenue, this is how the process of science is supposed to work, and generally it does very well.

Now back to Gilbert. He had been informed of a bowl-shaped depression in northern Arizona, with a rim elevated far above the surrounding limestone and sandstone plain; at that stage the formation was variously known as Coon Butte, Franklin's Hole, or Crater Mountain. Such a structure might be expected to result from a volcanic eruption, but the limestone and sandstone did not fit with such an idea: No basalts, granites, or other igneous rocks were to be found. What *was* found around the region, scattered as small shards within and outside the crater, were nickel-iron fragments similar in composition to many known meteorites. Gilbert was an excellent man in many respects, not the least of which was that he was willing to conceive of the history of the Earth as being linked with the history of the other objects in the solar system. In short, he was what in recent years has become known as a *planetary scientist*. Gilbert believed that the Earth was formed through the conglomeration of many smaller bodies drawn together by their mutual gravitational attraction. This is the central tenet of most modern cosmogonic theories. In this picture, after the planet agglomerated, a quiescent phase began, continuing to the present. However, Gilbert reasoned that an occasional straggler might arrive, late to add its mass to that of its many siblings which had accumulated in the early days of the Earth's history. This was an entirely reasonable idea; in fact, it is broadly correct.

As Ursula Marvin (of the Harvard-Smithsonian Center for Astrophysics) has eloquently pointed out, Gilbert liked the idea of a meteorite impact producing Coon Butte (later Meteor Crater) because it would explain three things: the crater, the meteorites, and the association of the two. Clearly, such structures are not commonly found around the globe, nor are concentrated fields of meteorite fragments, and so to believe that the two were commonly located by pure chance would be stretching credulity to the limit.

Would that be a problem for a scientist? Another central philosophy of scientific investigations is known as "Occam's Razor," (after William of Occam, a fourteenth-century English philosopher). Occam's Razor is a principle that states that entities must not be unnecessarily multiplied, so that any hypothesis must be as simple as is possible in order to agree with all the facts as they are known. This is sometimes paraphrased to merely say that the simplest solution is most likely to be the correct one. In the case of Meteor Crater, an impact origin would explain the crater and the meteorites; the problem for Gilbert, as we will see, is that an impact would not fit with other "facts," which he erroneously believed to be true and which led him to reject his impact hypothesis. Gilbert's credulity *was* stretched to the limit, and did not break, because of two other features that he wrongly thought to be characteristic of impacts.

The first was that the crater floor must conceal a huge iron meteorite, with the shards visibly scattered around being only a tiny fraction of the lode buried beneath. This is the same mistake that Daniel Barringer made when he bought the crater some years later, with the idea that a fortune was to be made from the iron ore he was sure would be found hidden below. It is obvious to us now that the majority of the impactor would vaporize on impact, the tiny meteorite fragments found being almost all that was left of the incoming object, but this was not recognized by Gilbert. So, when he looked for a magnetic anomaly over the crater floor, which he assumed would result from a massive buried iron meteorite, he was surprised to find that no detectable anomaly existed, telling him that no such ore body was concealed beneath his feet. The Canyon Diablo meteorite, found a few kilometers away, is apparently a substantial fragment of the impactor, but there is nothing but tiny shards of nickel-iron left in the vicinity of Meteor Crater itself.

Gilbert's second erroneous assumption was that an oval-shaped crater would be produced by an impact, whereas Meteor Crater is near-circular. In fact, it is somewhat polygonal in shape, reflecting the strengths of the rock formations from which it was wrenched; but it is certainly not elliptical. Gilbert reasoned that a circular crater could only result from a vertical impact, which is very unlikely,[6] so the crater must have had a terrestrial rather than an extraterrestrial origin. The reason that he was incorrect on this front is that he did not appreciate that it is energy release, not momentum deposition, that is important in such cryptoexplosive events as impacts.[7] Energy is a scalar, not a vector, so that the direction of arrival of the projectile is largely unimportant; the majority of craters on the surfaces of the Earth, the Moon, Mars, Mercury, and elsewhere in the solar system are near-circular. This is not to say that the idea of craters on other planets and satellites was one that was common prior to the acquisition of the spacecraft images that showed them in all their splendor. In fact, the intense cratering found on all old surfaces throughout the solar system seems to have been a series of surprises to many of the planetary scientists concerned, with few having considered the elementary calculations like those of Ernst Öpik in the 1950s. Öpik[8] satisfied himself that, at such a time as a spacecraft was sent to Mars, it would send back pictures of a densely cratered surface. That was indeed what was seen, with the "surprise" being repeated at Mercury, and then the Galilean satellites of Jupiter, and so on. Recently people have been surprised again by the craters mapped through the dense clouds of Venus by the high-resolution radar on board the *Magellan* spacecraft, the earlier Pioneer Venus radar having been unable to resolve the surface so well. Craters are everywhere that a solid surface exists to record impacts, ancient and modern.

What about Gilbert's expectation of an oval crater? Such a shape would be expected to result only from an impact at a very low angle, and though in the past few years there has been a claim that such a structure exists in Argentina, this idea is still disputed. The crater field in Argentina consists of a number of aligned ellipsoidal craters spread in an array 30 kilometers long and 2 kilometers wide. If this complex is proven to be impact-generated, it is also significant because of its youth, the geological evidence pointing to an age of less than 10,000 years.

But back to the last century, and to Gilbert's deliberations. Because there was no iron lode under the crater floor, and in any case the crater was not elliptical, Gilbert decided that Coon Butte/Meteor Crater could not have had an impact genesis; he suggested instead that it had been formed by an explosion of a huge volume of steam generated by the hot magma from the nearby volcanic field (the San Francisco mountains, just to the east of Flagstaff, Arizona), with the later arrival of a smattering of small meteoritic fragments being purely coincidental.

This false deduction, based on the correct scientific process but incorrect assumptions concerning the results of an impact, led to an abrupt halt in progress with regard to recognizing the importance of impacts in shaping the face of the Earth, and also the hazard that they pose to humankind. In the 1890s, Gilbert did not realize that in any hypervelocity impact, the projectile would be destroyed, circular craters resulting for the vast majority of arrival angles. Gilbert was loathe to relinquish his impact hypothesis, and his report mentioned ways in which the lack of a magnetic anomaly might be explained and the crater still be formed by an impact. For example, if the impactor was largely stoney, the iron shards found being atypical of its composition as a whole. Nevertheless, he was forced, by the state of knowledge in his era, to adopt his volcanic idea as the most likely explanation for the crater. Due to the position and prestige that Gilbert enjoyed, this was to result in an impermeable interpretation being adopted by the geological establishment in the United States, in particular by the USGS. Any dissenting view would face a hard fight for acceptance in the several decades to come. It was not until the 1950s that a new generation of geologists, Gene Shoemaker preeminent among them, would recognize the significance of Meteor Crater.[9]

As mentioned, in many respects Gilbert was a pioneer planetary scientist rather than just a geologist (in as much as that term implies exclusive study of the Earth). In 1892, he turned his attention to selenology, studying the lunar craters and basins by telescope. Because these were clearly quite different to terrestrial volcanic craters, he rejected igneous action as their origin. He was again tripped up, however, by his assumption that circular craters can only be produced by vertical impacts, and so developed a bizarre theory for their origin involving a ring of moonlets which coalesced to form the Moon proper, with various tilts leading to vertical impacts and therefore circular craters.

In was not until the middle of the first decade of the present century that Meteor Crater was to be correctly interpreted by Daniel Barringer and by Benjamin Tilghman, but their contributions were mostly ignored by the mainstream geological community. George Merrill of the Smithsonian Institution explained the total destruction of projectiles in hypervelocity impacts in 1908, making it possible for Gilbert's earlier ideas to perhaps be reconsidered and a reappraisal of the impact hypothesis made by members of the USGS. But Gilbert remained silent, and the mainstream took their cue from him. The study of crater-forming mechanics was held back for decades.

It was in the late 1920s that forward progress began. First came the correct interpretation of the Odessa Crater in Texas. Next, in 1927, came the first scientific expedition to the Tunguska explosion site in Siberia. The pace of activity was hastening worldwide, at least among those who were able to avoid the geological dogma that dictated all terrestrial phenomena to be endogenous. In 1932, a cluster of more than a dozen small craters with associated iron meteorites was found at Henbury, near the Northern Territory–South Australia border in the arid center of the Australian continent. In Saudi Arabia, two proximate craters were found, surrounded by nickel-iron meteorites and with interior walls smattered with fused silica glass into which tiny iron spherules were incorporated. In 1936, John Boon and Claude Albritton reinvestigated several craters in the United States which had previously been ascribed an explosive volcanic origin, following Gilbert's interpretation of Meteor Crater, and they deduced that these craters were actually created by impacts. There was no doubt, now, that impacts by extraterrestrial objects produced craters on the Earth. The discovery in the same decade of the first Earth-crossing asteroid opened up another source for the projectiles, quite apart from the well-known comets. Such craters were exogenous, not endogenous. The evidence was there to fuel a revolution in the way in which the factors sculpting the face of the Earth were viewed and investigated: It was just that almost half a century was to elapse before that revolution finally ran its course and impact cratering on the Earth was to be properly recognized as an important geologic process. Even today, there is still much to be done. For example, huge impacts have now been identified as occurring to some extent on a cycle similar to that of both mass faunal extinctions and various major geologic upheavals such as mountain-building phases. Could those impacts be responsible for the fracturing of crustal plates and thus the initiation of continental drift episodes? Are the two major revolutions of geological science in recent decades—that of plate tectonics (now largely completed) and of cosmic impacts (now underway)—directly related?

The four craters mentioned above are in northern Arizona, west Texas, the Arabian peninsula, and central Australia. These are areas characterized by their aridity, pointing to an important difference between the Moon and the Earth: Craters are relatively swiftly eroded on Earth by rain, snow, and wind, whereas on the Moon they remain for eons until a new projectile happens to erase the scar left by some cousin countless millennia before.

The occurrence of this handful of craters in arid regions gave an important clue as to where to look for more, but there is another fundamental difference between the Earth and its neighbor: Geology is active, but selenology is static. On the Earth, there are many volcanic regions spewing forth lava, whereas on the Moon, the interior cooled and solidified billions of years ago. Those terrestrial volcanoes tend to occur where huge crustal plates are splitting, such as the Rim of Fire around the Pacific or in the centers of oceans where the seafloor spreads and upwells, as in Hawaii and Iceland. What chance has a crater to survive for more than a few brief millions of years in regions such as the Andes, the European Alps, or even the Southern Alps of New Zealand if the continual buckling of the Earth's crust leads to their being crumpled and folded?

The places to look for craters, then, are the regions where there has been geological stability for hundreds of millions of years and where the climate is such that smaller craters are not quickly eroded away or indeed overgrown by vegetation, making their recognition unlikely. Would Meteor Crater have been found if the impact had been in the Rockies? Would it have resisted erosion for long if it had been in monsoon country?

The length of time that a crater will survive depends not only on the region in which it is formed, but also on its size. Obviously a crater 50 kilometers across will last longer than a comparative pipsqueak just 100 meters in diameter. For example, in Australia there are 19 known craters—probably there are dozens more awaiting discovery—and of those, four are less than about 6,000 years old. Those are the four smallest, however, all being less than 200 meters in diameter. Figure 8 shows vividly how small craters are all young, since they are rapidly lost through erosion. Meteor Crater is about 50,000 years old,[10] barely discernable as a tick on the clock that shows the geological and astronomical eons passing, and in a million years, maybe a little more, nothing will be left of one of the two great holes in the ground in northern Arizona; perhaps the Grand Canyon will survive for longer.

Early crater searches were based purely on the morphology of the structure: bowl-shaped depressions with raised rims, evidence of strata being peeled back, central uplifts in the larger formations as seen in many lunar craters. Craters are generally divided into two different classifications: simple and complex. Simple craters are bowl-shaped with rims raised above the surrounding terrain, as in the case of Meteor Crater. The maximum size of simple craters is about 2 to 4 kilometers on Earth,

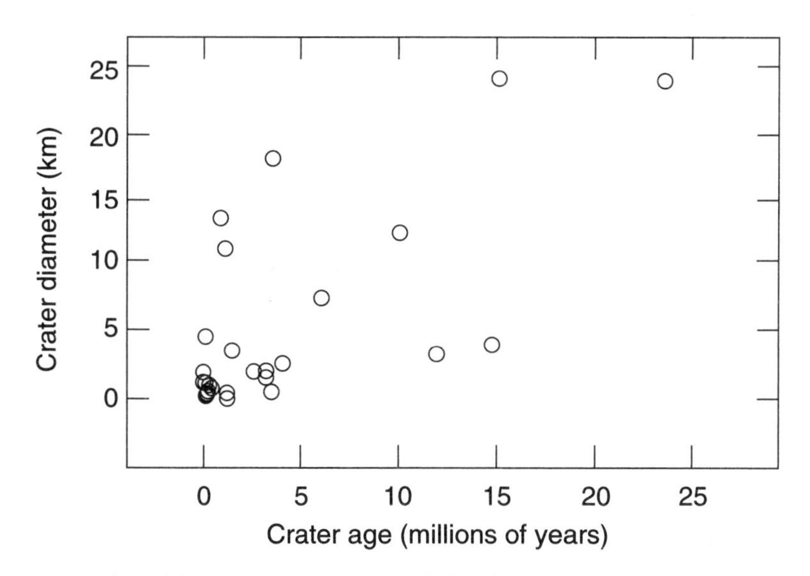

Figure 8. A plot of diameter against age for all dated terrestrial craters less than 25 million years old. It is not by accident that there is a concentration of younger, smaller craters: Only large craters survive erosion for more than a few million years. The disappearance of myriad smaller craters, which would otherwise pockmark the Earth, leaves us with a false sense of security.

the limit depending on a number of factors, such as the target rock strength. Typically, simple craters are five to ten times as wide as they are deep. Complex craters, with diameters of 4 kilometers and more, have larger diameter–to–depth ratios, often being a hundred times as wide as they are deep. Such craters may have central uplifts formed by the rebound of the target rock after impact, while it is still behaving like a fluid, and rims that have slumped so as to form a series of concentric rings.

These characteristics are pertinent for a relatively recently formed crater, but as weathering and other effects wear it down, pure morphological study is insufficient to allow an ancient impact structure to be recognized. The discovery of shock metamorphism produced in impacts, however, meant that a new tool could be applied: Even if the crater as such had been almost totally eroded away, still the existence of characteristic features in the target rocks would allow an impact to be established. In particular, *shatter cones*, conical striations produced in the target rocks by the extreme transient pressures in an impact, are important. These peculiar structures are now regarded as proof-positive[11] of the impact origin of a crater, whether heavily eroded or not. Another strong indicator is the presence of minerals that are only formed in very high pressure situations; in particular, the mineral known as stishovite forms from quartz only at very high pressures, but any mineral gradually produced deeply below the surface would have reverted to normal quartz in its migration upward. Stishovite on the surface is therefore diagnostic of an impact, a catastrophic event. Equally well, the melted rocks produced have a number of characteristics that are only explicable on the basis of the extreme conditions in an impact.

There are many impact craters now known on Earth, and these are certainly not equably distributed over the whole globe (see Figure 9). Major concentrations occur in North America, Eastern Europe, and Australia, due both to the geological stability of these areas and the more active search programs that have been carried out there. Seek and ye shall find.

Although more than 130 of these craters have been identified, several more being added each year and with discoveries accelerating, if

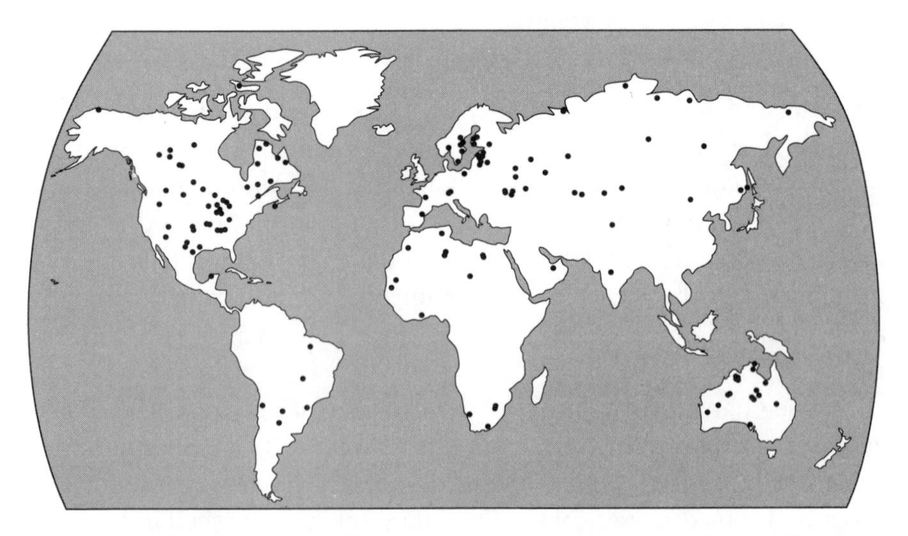

Figure 9. The geographical distribution of terrestrial impact craters, about 140 currently being known. Many hundreds, likely thousands, await recognition. The concentrations on this map reflect the fact that certain regions are more stable geologically speaking, making crater preservation more likely. Also note that the discovered craters delineate the areas where searches have been made or where the local inhabitants are more likely to recognize them (as in Europe and North America). Crater identification in tropical regions is also hampered by the obscuration of dense plant growth. (Courtesy of Richard Grieve and Janice Smith, Geological Survey of Canada)

anything, we should be amazed at their sparsity when compared with the number on the Moon. Most known terrestrial craters are less than 200 million years old, and yet the influx of projectiles over the past 3.5 to 3.8 billion years is believed to have been largely constant (that is, since the late heavy bombardment noted earlier). It may well be that we have yet to discover more than 1% of the impact structures on Earth. For example, the existence in Australia of four craters smaller than 200 meters in diameter and aged less than about 6,000 years implies that over the past 3 billion years or so at least 2 million such craters have been formed in Australia alone. Because these four probably represent, at most, 10% to 20% of the total still in existence, the actual number of these small impacts on Australia is likely to be in the tens of millions; and because the area of that island is just less than 8 million square kilometers, this means more than one impact per square kilometer. But wait: All such small craters that we have found were produced by iron bolides, and we know that these constitute a small fraction of all the projectiles, the stoney or cometary bodies blowing up in the atmosphere (see Chapter 9). This means that there have been many cosmic projectiles blowing up above or cratering each square kilometer. To look at it another way, if the 200,000 tons per year being accumulated by the Earth in this era has been more or less constant over the past 3.5 billion years, then each square meter has accumulated about a ton of cosmic debris.

Have all the bigger craters been found? The answer is no, far from it. Hundreds of craters are undoubtedly still hidden beneath the forest canopy of the Amazon basin, the tundra of the Arctic regions, the 70% of the Earth covered by water, and the shifting sands of northern Africa and Arabia. As mentioned in Chapter 1, a 180-kilometer-plus crater that is believed to be connected with the extinction of the dinosaurs was only recognized in 1990 in the Yucatán peninsula of Mexico. So far, only one submarine crater has been found, the 60-kilometer-wide, 50-million-year-old Montagnais structure in the coastal waters of Nova Scotia. It has been estimated that only about 10% of the craters larger than 10 kilometers in diameter and younger than 100 million years have been identified. Although many such craters may be lost from the record within a few million years due to erosion, shifting crustal plates, volcanism, or being buried under sedimentation, 10-kilometer craters are produced by impacts at a rate of about 10 per million years;[12] there must be many awaiting discovery.

Even recently formed craters can still go unnoticed, at least in the scientific literature and even in populated regions. Here I use the word *recently* in terms of a human lifetime, not as in geologically recent or astronomically recent. As an example, on the first day of June in 1937, a brilliant fireball was seen over Estonia, and a crater 8.5 meters in diameter was formed as the meteorite smashed into the ground. But this event has only undergone proper study in the past few years, over half a century later. Calculations of the original mass of the object, based on its brightness, put it above 50 tons, whereas the meteorite reaching the ground possessed only about one part in 700 of this mass, at most 67 kg, the rest having been burned off in the transit through the atmosphere and an explosion at a height of 28 kilometers. The Tunguska object, with a mass a thousand times 50 tons, produced no crater at all. Clearly, the atmosphere performs an excellent job of shielding us from the majority of the small projectiles that could wreak havoc if they landed intact. But has this shield made us dangerously blasé?

6

The Story of the Stones

We have seen that there is plentiful evidence from the craters found on the Earth to prove impacts by asteroids and comets in the past, and we have discussed how they might have caused widespread devastation, altered the climate, or detrimentally affected the biosphere in other ways. Apart from a few brief mentions of what is believed to have happened at the K/T boundary, however, we have yet to discuss the evidence that impacts have caused mass extinctions.

The first thing to note here is that the paleontological record can tell us little about anything except for the very largest catastrophes, such as that which appears to have happened at the K/T boundary about 65 million years ago. Since the time of that event, we expect that hundreds of impacts by 1-kilometer asteroids have occurred. The craters formed by some of these have been identified, but the fossil record is unlikely to tell us much about these impacts because they did not all cause mass extinctions. We have previously defined a "global event" as being an impact, the result of which is the loss of a full season's agricultural crop, or the occurrence of an environmental catastrophe similar to nuclear winter, or the downfall of civilization, with the deaths of

25% of humankind. Note that a 25% drop in the population of homo sapiens would not be noticed by future paleontologists: Fossil studies show only that a certain species existed until a certain horizon and not thereafter. They reveal only actual extinctions, where 100% of a species, or of several species, disappears. In terms of the long-term future of humankind, a seemingly apocalyptic event that killed 25% of us might actually be of little consequence. The population drop might well be recovered within a few generations in the same way that the mass killings in wars are soon numerically replenished. Indeed, humankind might positively flourish as a result of an impact, in the same way that technological breakthroughs are often made under the stringencies of warfare (although it would, of course, be preferable if such conflicts were not necessary).

The point here is that we are concerned not only with cataclysmic impacts but also especially with impacts that, were they to happen today, would threaten the end of civilization as we know it. As argued earlier, these are events that can be expected to occur about once every 100,000 years and perhaps a lot more frequently. Although it is most unlikely that such a hazard will face us again soon, it is inevitable that one day it will, if humankind survives for long enough. But more to the point here, we do need to consider the story in the stones. We need to evaluate what paleontology and geology can tell us about what occurred in past eons. This will inform us not only about the link between impacts and mass extinctions (and thus about how evolution proceeds), but also about the role of such impacts in controlling how the face of the Earth is shaped.

We can look at the effects of impacts on all scales and in all time frames. Starting with the longest time frame, stretching back to about 3,800 million years ago—thus predating the appearance of life on Earth—links between gigantic impacts on the Moon and tectonic events on Earth (that is, mountain-forming drifting of crustal plates) have been found. Richard Stothers of the Goddard Institute for Space Studies in New York has found that the lunar cratering record of the past 3,800 million years shows six main episodes of bombardment, and that (to within the accuracy of the dating) these coincide with the six main periods of mountain-building activity on Earth. Although this has yet to be confirmed by other studies and more precise dating, it does seem to indicate that on the grandest scale of geological events—the break-up of continental plates on Earth—asteroid and comet impacts are implicated. What we are predominantly interested in here, however, is how impacts might affect life, in our future and in the Earth's past. Archaic impacts might have limited the early evolution of life, but of more direct interest is how such impacts may have controlled evolution's progress over the past 500 or 600 million years. This is also the period over which we might reasonably expect large craters to be preserved, allowing us to date them and to look for correlations with mass extinctions.

The peculiar thing about crater ages, though, is that they do not appear to be random, but instead are periodic, as shown in Figure 10. That is, there are distinct epochs in which impacts occurred more frequently, interspersed by eras of comparative quiescence. Although there has been much debate over this, because there are only a few dozen reasonably well dated craters, it now seems fairly accurate that cratering is periodic, at least to some extent. For example, in another paper, Stothers considered 23 craters larger than ten kilometers in diameter and less than 250 million years old and found the signs of a periodicity of about 30 million years. It is not possible to be precise because the individual dating errors are about 6 million years; others have derived a periodicity closer to 26 million years with basically the same set of data.

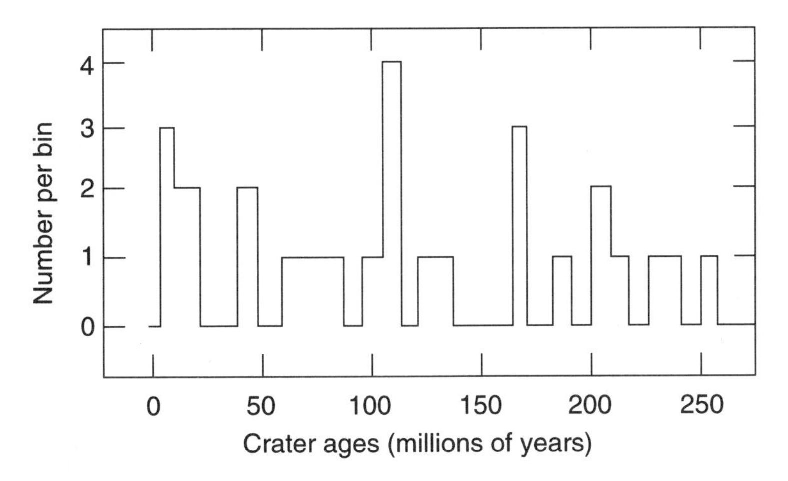

Figure 10. Histogram of the ages of terrestrial impact craters younger than 260 million years and larger than 5 kilometers in size, in 8-million-year bins. Even with the poor dating of many of these craters, this figure is suggestive of a 26-million-year periodicity.

Clearly, a larger set of crater ages is required in order to establish definitely whether there have been periods of high impact activity, and as time goes by more and more terrestrial craters are being identified and dated, making a refinement possible. For example, Shin Yabushita of the University of Kyoto in Japan has analyzed the ages of 98 craters less than 600 million years old and has found that a periodicity of 30 million years is demonstrated by those that are *less* than 10 kilometers in diameter (57 craters), but that the periodicity is not so clear for those that are larger (41 craters). This does not mean that the larger craters are proven *not* to be periodic; it just means that we need more and better data to make any progress.

This does not mean that *all* craters are formed only in episodes of high activity once every 30 million years, but that there is a background influx to the Earth (from asteroids and comets such as those that we observe currently), but with an overlying increase in the impact rate from time to time. There have been a number of suggestions as to how such increases might come about, most of which have since been shown to be in error. One idea, due to American physicists David Whitmire and John Matese, was that there is a massive Planet X circuiting the solar system. Although this planet would normally have an orbital period of only some centuries, Whitmore and Matese pictured it in an eccentric orbit—in the astronomical sense, meaning *noncircular*—which precesses (swivels around) such that every 30 million years it would encounter the swarm of comets thought to inhabit the region that ranges from 50 out to a few hundred astronomical units from the Sun.[1] Some of these comets would then be diverted by Planet X into orbits entering the inner solar system, striking our planet occasionally. However, any such planet would also disrupt the orbits of the outer planets to a degree that is proven not to have occurred by our measurements of their movements. Also mitigating against the theory is the fact that such an orbit for Planet X would itself be unstable over a length of time equivalent to the age of the solar system.

So what other explanations might account for cometary waves causing periodic impacts on the Earth every 30 million years or so? A variant on the Planet X theory of Whitmire and Matese was the Nemesis hypothesis, which holds that the Sun is part of a binary star system. The idea is that the Sun has a companion star, a brown dwarf that would be so faint that it would so far have escaped telescopic discovery. This hypothetical star has been nicknamed Nemesis, the death star. It would have an eccentric orbit and presently be near its furthest point from the Sun and the Earth, over a light-year distant, making its nondiscovery unsurprising. Once every 26 to 30 million years (this being its orbital period around the Sun), however, it would plunge in closer, through the Oort cloud, scattering comets as it did so into all sorts of new trajectories. Some of these would end up crossing the Earth's orbit. A comet shower would be produced, the flux of comets in the inner solar system being enhanced by a substantial factor for a few million years. An increased impact rate on the Earth (and other planets) would be anticipated, introducing a periodic component into the age distribution of impact craters.

The Nemesis theory was a nice idea. So nice that it was independently invented by two teams of scientists: Richard Muller, Marc Davis, and Piet Hut (of the University of California at Berkeley) and David Whitmire (University of Southwest Louisiana) and Al Jackson (Lockheed Corporation, Houston). Obviously Whitmire had his thumbs in two different pies here (an approach that I applaud; too many scientists become wedded to favorite theories and refuse to budge, even against mounting contrary evidence). Unfortunately, nice ideas are not always the correct ones, and the instability of such an orbit as that suggested for Nemesis has proven to be a difficulty, although the Berkeley team continues a telescopic search for it.

So, what could cause the apparent periodic influx of impactors to the Earth? The only mechanism[2] seems to be a periodic disturbance of the Oort cloud leading to a comet shower, the question being the source of that disturbance. Back in 1979, Bill Napier and Victor Clube (both then at The Royal Observatory, Edinburgh, but later at the University of Oxford) had suggested that catastrophic impacts on the Earth could result from Oort cloud–derived comet showers produced every so often as the Sun orbited the center of the galaxy, some of the comets in the cloud being knocked from their regular orbits as the solar system passed through the galactic spiral arms. This was the gist of a theory that seems to be the correct one (or at least the favorite at the moment), although Napier and Clube were incorrect in that they were considering the movement of the Sun in the galactic plane to be the most significant effect. There was a good reason for this assumption: Lunar rocks returned by the Apollo program had shown that the micrometeorite flux onto the Moon follows a 250-million-year periodicity, which is similar to the orbital period of the Sun about the galactic center. Napier and Clube therefore reasoned that either small particles cloaking our atmosphere, or larger impactors arriving at the same times, could cause environmental catastrophes on Earth. Although the gist of their idea seems to have been correct, it was a few years before it was realized that it is the solar motion *perpendicular* to the galactic plane, rather than parallel, that is the significant factor.

By the mid-1980s, it was becoming apparent that crater ages were perhaps periodic. This realization was achieved independently by Walter Alvarez and Richard Muller at Berkeley and by Michael Rampino and Richard Stothers in New York. The Nemesis hypothesis for the periodicity followed from the Berkeley physicists, while Rampino and Stothers came up with another suggestion. It was already known that the Sun, in its orbit about the galactic center, does not remain in the plane of the galaxy but concurrently oscillates up and down through that plane. It takes around 60 million years to execute each complete oscillation, meaning that it passes through the galactic plane once every 30 million years. If the Oort cloud were perturbed in some way each time the solar system passed through the more densely populated galactic plane, perhaps comet showers would be produced.

But what would cause the Oort cloud perturbations? If one goes out on a clear night, one sees the Milky Way, formed by the many billions of stars that are concentrated toward the plane of our galaxy, stretched overhead. The first idea, then, was that passing stars could stir up the Oort cloud, with the solar system encountering many more stars than usual as it passed through the galactic plane. When the calculations were done, however, it turned out that such stellar encounters were not frequent nor close enough to throw sufficient comets out of the Oort cloud.[3]

The stars, however, are only the masses that our eyes see delineating the Milky Way. There are other structures present which are also potential perturbers, but are not so easily seen. In particular, we know that stars form when large clouds of gas and dust collapse under their own gravity, the centers heating and condensing until the central temperature is high enough to support the fusion of hydrogen nuclei so as to form helium, whereupon the cloud has become a star and glows in its own right. At that stage, the energy output leads to an expansion of the remnant gas and dust around the central condensation, it being blown outward. From this residual cloud, it is believed that planets may form. Stars have been observed condensing in this way, in one form of what astronomers call *nebulae*; when discussing the origin of the solar system, the cloud from which the Sun and planets are assumed to have formed is called the *pre-solar nebula*.

Throughout its life, a star emits a stream of particles that make their way out into interstellar space. In the case of the Sun, this stream is called the *solar wind*, and its interaction with our atmosphere can affect radio communications and the weather. After a star has gone through its lifetime, it may explode and recycle its constituent material, while other stars regularly throw off clumps of material. Most of the material of which the Earth (and each individual human) is composed has previously gone through at least one stellar cycle. For example, elements lighter than iron can only be produced in fusion reactions within stars, hydrogen nuclei being the original building blocks.

Interstellar space, then, is not empty but is populated by huge clouds of gas and dust which may be about to collapse to form a star or may yet be stable for billions of years. Using radio telescopes, astronomers are able to observe the basic properties of these structures, which are called *giant molecular clouds* (GMCs). Apparently it is these which provoke comet waves, since the solar system is much more likely to encounter one or more GMCs as it oscillates through the galactic plane than it is to have a near passage by a star. Although diffuse, these clouds are extremely large (making encounters more likely) and have masses far in excess of individual stars, meaning that they are capable of severely perturbing the orbits of comets on previously stable paths, throwing some onto trajectories passing through the inner solar system.

Before moving on to the induced cometary/asteroidal impacts on the Earth, it is interesting to note that GMCs have been implicated in terrestrial catastrophes in another way. Obviously, such structures, being light-years across, might not only pass by the solar system, stirring up the Oort cloud, but could envelop the planetary region itself. Under such circumstances, we might expect both the planetary and the interplanetary environments to be substantially perturbed. Although a GMC is diffuse, it would nevertheless affect the flow of the solar wind. It might also scatter enough light such that there is no real night-side of the Earth. Another possibility is rather more intriguing, however. The Earth could be enveloped in a cloud largely composed of hydrogen molecules, and for a period of millions of years it would gradually accumulate these. We all know that hydrogen is inflammable, so one would expect that this hydrogen would gradually be incorporated with the atmospheric oxygen, producing water. Yabushita, with a co-worker, British astronomer Tony Allen, has suggested that the passage of the solar system through such GMCs could severely perturb the terrestrial environment.

In the past two years, supporting evidence for this scenario has been identified. It has been found that the oxygen content of the atmosphere fell from about 35% to 28% in the 2 million years prior to the K/T boundary event. One explanation for this is that hydrogen from a GMC effectively consumed the oxygen. This would have two important implications. First, the time taken for comets to fall in from the Oort cloud is a few million years, so we may be seeing the signature of a GMC perturbing the Oort cloud such that impacts later occur, but in the meantime the GMC has directly affected the Earth itself.[4] Second, such a fall in oxygen, if it were due to combination with hydrogen so as to produce water, would result in the precipitation of about a meter's depth of water over the whole planet. This would clearly have a significant effect on the global climate.

We can now return to impacts on the Earth and their apparent periodicity. The identification of a periodicity of around 26 to 30 million years in the cratering record, as announced by Walter Alvarez and Richard Muller in mid-1984, is remarkable in itself (given that suitable data had been available for some years), but the most extraordinary thing was that it followed hard on the heels of the publication of a paper by David Raup and Jack Sepkoski (University of Chicago) indicating that mass extinctions of fauna also followed such a cycle (see Figures 11 and 12). That is, scientists were no longer dealing with a one-off event, such as had been suggested by Luis Alvarez in 1980 in collaboration with his son Walter, Frank Asaro, and Helen Michels. They identified an anomalous amount of the element iridium at the K/T boundary and provoked a storm when they hypothesized that this was deposited by the impact of a large asteroid 65 million years ago, resulting in the extinction of the dinosaurs and many other animal families. The game had been widened to include up to ten mass extinctions over the past 260 million years, these appearing to follow the same clockwork cycle as crater formation on the Earth. Even if the iridium layer were accepted as being

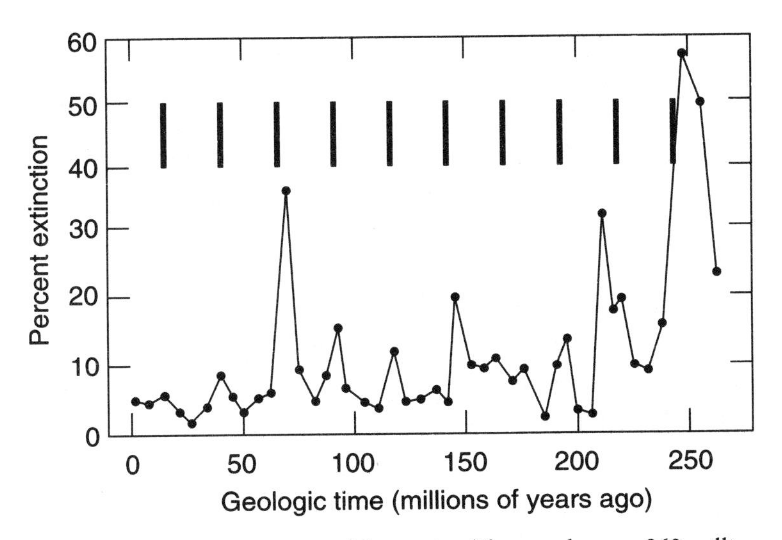

Figure 11. Percentage extinction record for marine life over the past 260 million years. The bars at the top delineate 26-million-year cycles, these correlating in nine cases out of the ten with an extinction event of some magnitude—and one could even argue for the tenth. (After J.J. Sepkoski, "The taxonomic structure of periodic extinctions," Geological Society of America, Special Paper 247, pp. 33–44, 1990)

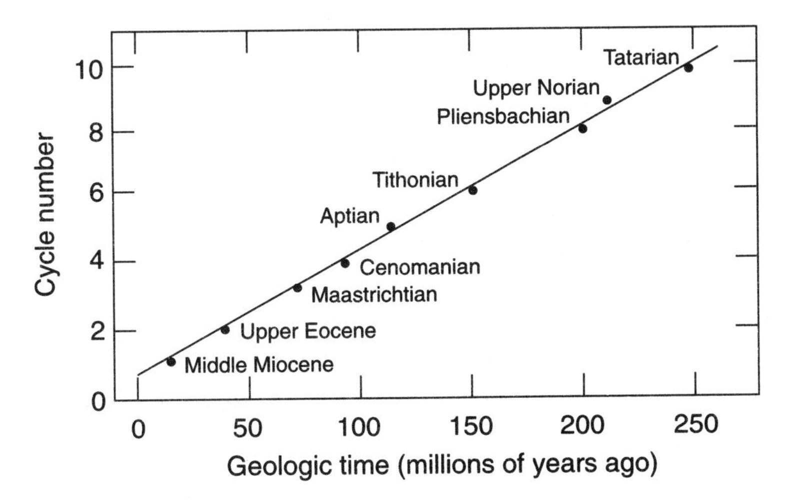

Figure 12. The dates of the nine mass extinctions in Figure 11, but plotted with a straight line. A mass extinction might be expected in the Middle Jurassic (between the Pliensbachian and Tithonian extinctions) but is not clearly defined in the presently available paleontological data, though there is a period of heightened extinction rate apparent in Figure 11 between 150 and 175 million years ago. The Maastrichtian extinction is that in which the dinosaurs disappeared.

due to the arrival of extraterrestrial material—and in the 1980s the Chicxulub Crater had yet to be identified with the K/T catastrophe, so that at that stage the iridium could still be interpreted as being deposited by a gradual influx of meteoroids and dust—it could still be argued that the synchronism of the iridium deposition and the mass extinction might be merely a coincidence. This argument could not possibly be used if there were several mass extinctions occurring at the same time (to within dating errors) of large impacts, especially when both phenomena appeared to be cyclic and following the same basic periodicity, indicating the hand of some astronomical pacemaker. There is no known terrestrial driving force that could explain a regular 26- to 30-million-year cycle.

Of course, the similarity in periodicity does not in itself prove the geological perturbations were caused by impacts. Terrestrial events, in particular various geological phenomena, have, after all, been shown to follow regular cycles of their own. Since the beginning of the twentieth century, there have been suggestions that geological events are also periodic in some fashion, and as early as 1921, the prominent American astronomer Harlow Shapley was suggesting that these were related in some way to the motion of the Sun about the galaxy. There is, in fact, a long history of ideas of geologic periodicity. Probably the most prominent person in recent years to investigate the common periodicity of geologic cycles has been Michael Rampino, who works at the Goddard Institute for Space Studies in New York City and at New York University. With Ken Caldeira of Pennsylvania State University, Rampino has studied major geologic events of the past 260 million years (the best-dated period, of course, covering the time over which mass extinction data, from Raup and Sepkoski, and others, are available). The results show that these events—among them, continent-wide flood basalt (volcanic) eruptions, sea-level changes, ocean anoxic events (that is, loss of oxygen from the oceans), rates of sea-floor spreading, mountain-building events, and major salt deposition—occur with a periodicity similar to the mass extinctions and the impact cratering.

So prevalent is the commonality of the periods found that Rampino has argued in favor of what is known as the "Shiva hypothesis." Many readers will be familiar with the Gaia hypothesis, which holds that the Earth functions like a giant organism, modulating any environmental perturbations so as to ensure that life can continue to proliferate. This idea is most often identified with the British scientist James Lovelock and American biologist Lynn Margulis. The Shiva hypothesis, alternatively, sees the Earth as being predominantly under the control of cosmic influences, needing to respond every so often as the planetary environment is shocked by extraterrestrial parameters. This idea is named for Shiva, the Hindu goddess of life and rebirth.

An obvious question that arises, if one accepts that periodic waves of comets and asteroids arrive in the inner solar system and impact the planets, including the Earth, is, "When is the next wave due?" Of course, what we need to ask is, "When was the last wave?" The answer is not easy to find. Dating of craters does show a large number of recently formed craters, but this could be due to their preferential survival to the present. Rampino's analysis of the other geologic phenomena results in a deduction that the last wave was, at most, 9 million years ago and, quite likely, much closer to the present. Yabushita's work on crater ages results in a conclusion that the last wave arrived within the past 1 to 5 million years. Astronomical observations of the motion of the solar system about the galaxy indicate that we passed through the galactic plane about 3 million years ago. These all indicate that a comet wave is not due for at least 15 million years, likely longer; a more important question is whether we are currently on the tail of a wave, such that the present cratering rate (and hence impact hazard) is higher than the long-term (billion-year) average.

This question has been the subject of some debate, and my own view is that it is still open for discussion and argument. Certainly the various phenomena investigated (craters, geologic events, mass extinctions) show a similar periodicity, but there are difficulties in showing that they are co-phasal. This might be due to problems in the accuracy of the dating (different methods, and the differing radioactivity timescales used, can lead to variations of millions of years in the individual dates and hence substantial phase errors), but there is also another possible explanation. Not all geologic phenomena follow a cycle of about 30 million years. Although there is still some debate on this front, it seems likely that reversals of the Earth's magnetic field have a period of around 15 million years, and in the last decade it has been suggested that these geomagnetic field-swaps are induced by impacts, although the exact physical mechanism is unclear. Thus some scientists have argued that the "true" period is in fact 15 million years, with the 30-million-year period more obvious due to every second cycle being stronger. Certainly the severities of the terrestrial catastrophes seem to be quite variable from cycle to cycle. Bill Napier and Victor Clube, in particular, have vociferously stressed that the 15-million-year periodicity is that which should be studied, pointing out that the phasing discrepancies between the disparate types of data tend to disappear if that cycle is adopted, rather than 30 million years, especially if only the most precise ages are used and the less well dated craters are excluded. In this respect, it is notable that Yabushita's analysis of crater ages results in a highest likelihood of about 16 to 17 million years being the fundamental period. Having stated that, it seems that a 15-million-year cycle would be at odds with the galactic plane crossing theory of cometary waves, requiring the inducement of four cometary waves per 60-million-year oscillation about the galactic plane. Obviously, there is still much work to be done in this intriguing area.

Although it seems conclusive to many in the field that asteroid and comet impacts are related to mass extinctions and boundary events, there are still some who cling to endogenous causes for these events. In particular, it has been repeatedly pointed out that at the time of the K/T event 65 million years ago, there were widespread volcanic eruptions to the west of what we now call India, a flood basalt area called the "Deccan traps." Similarly, there is a flood basalt area in Siberia which has been dated as being contemporaneous with the Late Permian mass extinction around 245 million years ago. These two extinction events are by far the most prominent in the past 300 million years in terms of percentages of species and families terminated. The idea is that global climate change, produced by the huge release of water vapor, sulfur dioxide, and other noxious gases from the eruptions, could have caused the extinctions.

While the impact lobby does not necessarily claim that the impacts were directly and solely responsible for the mass extinctions (refer to Table 3 and the previous discussion of geologic upheavals apparently related to impacts), it is clear that the root cause in each case is the influx of extraterrestrial material (of various dimensions) to the Earth. The question then is whether the volcanic eruptions could have been triggered by the impacts.

Although the taxpayer is often impatient with the amount of money spent on spacecraft sent to other planets, it is surprising how often such investigations help us understand phenomena here on Earth. Here is a good example. NASA's *Mariner 10* spacecraft is the only satellite ever to be sent to Mercury. In 1974–75, it looped around the Sun, having three fly-bys of the planet during which it was able to obtain images of about 50% of Mercury's surface. Humankind had never before been able to see details of the face of that planet because it is too far from the Earth and too near the Sun for telescopic study. It should not have been a surprise (although it was in some quarters) to find that Mercury is pockmarked with craters, looking broadly similar to our Moon, because, like the Moon, Mercury has no atmosphere to shield it from asteroids, comets, and meteoroids. In particular, one vast impact basin, now called Caloris, was identified. The funny thing was that on the opposite side of the planet, at the antipodal point to Caloris, the Mariner 10 scientists found that there was a region of peculiar "chaotic" terrain, looking as if the ground had been shaken up and shattered, even though there were no significant impact craters nearby. A plausible hypothesis for the origin of the chaotic terrain was soon invented: When Caloris was formed, huge seismic waves were focused through the interior of Mercury, meeting at the antipodal point and breaking up the smooth terrain that previously existed there.

How does this apply to the Earth? Although both the Chicxulub crater and the Deccan traps are now about 15 to 20 degrees north of the equator (India having moved substantially north since the eruptions), they are separated by close to 180 degrees in longitude. When the impact occurred, India was much further south, near the antipodal point to Chicxulub. The crust-puncturing giant impact in Mexico apparently induced seismic waves which were focused on western India, causing fracturing which then led to the widespread Deccan eruptions. These occurred episodically, and dinosaur fossils and eggshells have been found between the first and second sets of eruptions. This lends weight to the idea that, if the Deccan trap eruptions were due to seismic focusing of the type described here, then the K/T boundary extinctions were not due solely to the Chicxulub impact, which fits in with the idea of multiple impacts over a period of time as the cometary wave dispersed.

And what of the Siberian eruptions? Rampino has recently been looking for craters that might be linked to the Late Permian extinction, the most severe such event in the geologic record. The appropriate antipodal point to Siberia is the Falkland plateau off Tierra del Fuego in the far South Atlantic. On that plateau, Rampino has identified two sub-oceanic circular basins with diameters of about 350 kilometers and 200 kilometers, respectively. Dating of the rocks indicates the same age as the Late Permian extinction. Although this investigation is still in the earliest stage, it seems that the finger of suspicion may be pointed at these apparent impacts as responsible for the catastrophe that ended the Permian, inducing the volcanic eruptions in Siberia and the greatest mass killing in the past few hundred million years.

As a closing comment to this chapter, the following is pertinent. In recent years, it could be imagined from the media coverage (and indeed many scientific papers) that Luis Alvarez and co-workers "invented" the idea of an impact causing the extinction of the dinosaurs, which is far from the case. This is not to decry the Alvarez team's work, which was seminal; it provided the first hard evidence that such an impact actually occurred. By 1980, however, it was an idea already some decades old (and much more than that if we trace things back to Sir Edmond Halley, as described in Chapter 2). Following the discovery in the 1930s of the first three Earth-crossing asteroids, the great American meteoriticist Harvey Nininger recognized that impacts by such objects must occur from time to time with calamitous consequences. We have already discussed (also in Chapter 2) the early work of Ralph Baldwin and Fletcher Watson, in which they set out the implications of these asteroid discoveries for the lunar and terrestrial cratering records. But Nininger went beyond that. In 1942, he published a brief report in which he suggested that geological boundary events and mass extinctions might be due to such impacts, although he did not specifically mention the dinosaurs. In 1956, paleontologist M.W. De Laubenfels (Oregon State University) appears to have been the first to link dinosaur extinctions with impacts, taking his lead from the Tunguska event of 1908 and scaling the explosions up. In 1973, Harold Urey, the late American geochemist who over many decades made several seminal contributions to planetary science while working at the University of Chicago and the University of California at San Diego, showed that certain fragments of melted rock, called *tektites*,[5] have radio-dating ages similar to the last few geological boundary events, suggesting that the boundaries were caused by massive comet impacts. In 1979, a year before the Alvarez team's paper was published, Napier and Clube pointed out that the ages of several impact craters formed in the past 400 million years seemed to be coincidental with geological boundaries and mass extinctions. In the same year, U.S. geologists C.K. Seyfert and L.A. Sirkin had suggested that enormous impacts could be responsible for the break-up of crustal plates and the stimulation of plate tectonic movement, heralding new geologic eras. The Alvarez group's paper did not, then, come out of the blue; it was a major step forward, but not the first step. The media ballyhoo is all part of the ways of science—provoking scientists in disparate areas to start to take the large impact theory seriously. It is only now that we can look back and consider the true history of the concept.

Ideas of the relationship between massive impacts on the Earth and mass extinctions have come a long way over the past 10 or 15 years. The idea of "punctuated equilibria" being the mode through which the evolution of life forms mainly takes place, rather than the gradual changes visualized by Charles Darwin, was promoted by Niles Eldredge and Stephen Jay Gould in the early 1970s (although it was first suggested more than a century earlier). Advances in our understanding of the influence of impacts are now showing that the equilibria may be "punctured" rather than just punctuated, the puncturing agents being asteroids and comets that every so often invade our territory, wreaking havoc as they do so.

7

Coherent Catastrophism

In the previous chapter we saw that impacts on the Earth do not occur randomly in time, with there being strong evidence to suggest that at least every 30 million years or so there occurs a wave of comets, producing a period in which there are more impacts than the long-term average. This means that impacts are not random on time scales of millions of years, though that is of little interest to humankind with regard to our self-preservation. We would be interested, however, if impacts were non-random on time scales such as centuries or millennia, because that would imply that there are dangerous and not-so-dangerous epochs spaced by periods not much more than a handful of generations.

In discussing the impact rate on the Earth, the assumption is almost always made that these impacts do occur randomly in time, with little or no thought being given to the validity of that assumption. We have discussed how Luis Alvarez, in concert with his son Walter and others, uncovered important evidence suggesting a link between an asteroid impact and the mass extinction at the end of the Cretaceous era; concerning another important publication by Alvarez, on a totally different topic, Richard Garwin (of the IBM Thomas J. Watson Research Center in New York) wrote

> This paper is rich in its informal advice for the physicist, or for anyone wanting to arrive at the truth or to persuade others . . . The physicist *always* asks, "Do I know this?", "How do I know this?", and "Is this still true?"

This lesson, eloquently put, seems to have been forgotten (or never learned) by many in the impact field, physicists or not, with very little questioning of the basic assumptions taking place. It is certain that the influx of small particles to the Earth is highly time-dependent; whether this is also the case for larger particles is the question that must be addressed. To understand the possible time variation in large impacts, we must first understand how the smaller particles—meteoroids—behave. So, let us have a short tutorial.

Find a clear dark site one night and let your eyes become well adjusted to the low light levels. Then, by looking randomly across the sky, you may expect to see up to ten meteors (or shooting stars) an hour. In fact, most people are asleep during the best time for viewing meteors (during the last hour or two before dawn), when the count rate is two or three times higher than in the late evening. Radar observations of meteors—which are not limited by either daylight or cloud cover—show that from any observation site, the number of meteors detected varies sinusoidally, with a peak at 6 A.M. and a minimum at 6 P.M.

This is easily understood if you think of the Earth as being a large body continually passing through a cloud of smaller objects (meteoroids), all moving in different directions: The Earth runs into more meteoroids on its leading face (the apex), and you are central on that leading face every day at 6 A.M. Conversely, for any meteoroid to hit the Earth at the 6 P.M. position, it has to catch up from behind.

The five or ten meteors that you might see in an hour will arrive from different directions; they are part of what is called the *sporadic background*. Often people say that sporadic meteors come from random directions, but this is incorrect. In fact, their arrival directions might be said to be semi-random in that, although you may see no pattern in their right ascensions of origin (the astronomical equivalent of geographical longitude), nevertheless, most of them have radiants close to the ecliptic, meaning that they generally have low radiant declinations (the astronomical equivalent of geographical latitude). In fact, the declination of the ecliptic, along which the meteors are clustered, varies during the year, which is fairly obvious when you recall that the ecliptic is also the path followed by the Sun across the sky, with the seasons depending on its declination. Summer occurs in the northern hemisphere when the Sun has come as far north as it ever does. This also implies that there is a seasonal variation in meteor counts, because on the dates that the ecliptic's declination is closest to your latitude, you are closest to being on the leading point of the Earth on its path around the Sun.

When the original orbits of meteoroids are determined, through various radar or optical techniques, it is found that most sporadic meteors have orbits basically similar to Earth-crossing asteroids: small inclinations to the ecliptic (and therefore those low-declination radiants), short-period orbits, and aphelia closer than Jupiter. Although their orbits are not really random, sporadic meteors are literally sporadic in that they arrive randomly in time. I have said that you might see five to ten an hour, but some hours you might see 20 scattered during that hour, but in the next hour only three or four.

Every so often when you go out meteor-spotting, however, you will see something rather different. On certain nights of the year, a larger number of meteors will be seen, and although the sporadic meteors still come plodding in at five or ten an hour, perhaps an additional 50 an hour will be observed, all seeming to emanate from the same region of the sky. Such an occurrence is called a *meteor shower*, and the common area of emanation is called the *shower radiant*. Most such showers occur annually on the dates that the Earth passes through a meteoroid stream, that being a loop of meteoroids that has formed through their release from some much larger parent object—an asteroid or a comet. Each shower is usually distinguished by the constellation from which it appears to come—for example, the *Geminid* shower, or the *Orionid* shower—although at times the existence of more than one shower (at different times of year) radiating from any particular constellation will lead to each being labeled according to the star that the radiant is nearest; examples are the Eta Aquarids, the Iota Aquarids, and the Delta

Aquarids, which are not genetically related. The meteoroids do not actually originate from that star or constellation; perspective effects just make them appear to do so.

Because it takes about a week to pass through a stream, showers will continue to show pronounced activity for some days. There may be no significant variations in count rates from night to night, as is the case for the Eta Aquarids in the first week of May, or there may be marked rate changes, as is the case for the Geminids, for which activity gradually builds in early December, reaching a peak around December 13 or 14 with a rapid fall-off thereafter.

Many well-known showers have been linked with specific comets. For example, we know that the Eta Aquarids and the Orionids (in the third week of October) are meteoroids that originated from Comet Halley. P/Halley produces two showers because the Earth passes close by its orbit twice each year (that is, once on the pre-perihelion leg of the orbit and once on the post-perihelion leg). Because the stream is a complete loop around the cometary orbit, the showers are seen annually. Thus if you missed seeing Comet Halley in 1986, at least you can see some small parts of it every year in May and October. For other streams, the loop is not necessarily uniform, because the meteoroids take some time to be dispersed from their parent. Thus the rising activity of the Perseid shower beginning in the late 1980s forewarned astronomers that the parent comet, P/Swift-Tuttle, was about to reappear, as it duly did in 1992; recall that this is the comet we discussed in Chapter 1.

Prior to the past decade, it was believed that only comets spawned meteoroid streams. In 1983, however, data returned by the Infra-Red Astronomy Satellite led to the discovery of asteroid 3200 Phaethon, which was soon recognized by Fred Whipple to be in an orbit indistinguishable from the Geminid stream. Whipple, of the Harvard-Smithsonian Center for Astrophysics, is undoubtedly the doyen of astronomers working in this area, having made fundamental discoveries and developed important understandings of comets and meteoroids from the 1930s through to the present. Later in the 1980s, using meteor orbits measured in Australia along with the catalogue of Earth-crossing asteroids as it stood then, I showed that there were several other asteroids that could be tied to meteoroid streams. This work, which is continuing, seems to indicate that these asteroids were once active comets, but have since either become totally devolatilized or perhaps just dormant in the present epoch. Obviously, the association of Earth-crossing asteroids with meteor showers is an important pursuit because it gives us a clue to the origin of at least some of those asteroids: defunct comets, as opposed to main-belt asteroids that have been diverted into planet-crossing orbits.

The observation of meteor showers has an even more important implication vis-a-vis asteroid and comet impacts on the Earth, however. Although sporadic meteors arrive at random times, shower meteors do not. I can state with certainty that on December 13 of each year, millions of Geminid meteoroids will enter the atmosphere, while around August 12, myriad Perseid meteors will light up the sky. Indeed, the Perseids are sometimes called *The Tears of St. Lawrence*, because in the past they peaked a couple of days earlier, on the feast day of that saint. It is therefore clear that the influx of extraterrestrial material to the Earth does not occur randomly in time. But how large is that mass flux, and in what sized particles is it concentrated?

On the average, the Earth accumulates from interplanetary space each year approximately 200,000 tons of solid material. Of this, the majority comes in the form of discrete massive impactors. For example, we have seen that a 1-kilometer asteroid arrives about once every 100,000 years; because such an object has a mass of about 1.5 billion tons, over the long term such bodies contribute around 15,000 tons per annum to the terrestrial influx. Summing the mean influx for all macroscopic objects, from 1-meter meteoroids to the largest feasible comet (that is, effectively adding up the mass influx under the curve to the right of Figure 6), one derives a rough estimate of about 160,000 tons per year, but it is important to understand that this arrives spasmodically, and mostly in large lumps. The overall mass influx is dominated by the large objects (asteroids and comets) that hit infrequently. About 20% of the annual influx arrives more or less continually in the form of particles from the dimensions of dust grains (which decelerate in the atmosphere and eventually settle to the ground) to 1-ton meteoroids (up to about a meter across, producing very bright fireballs). This amounts to about 40,000 tons (the total under the curve to the left of Figure 6), or around 110 tons per day, although the influx does vary considerably during the year due to the occurrence of meteor showers, as we are discussing here.

Of that 110 tons a day, the majority is in the form of meteoroids that produce meteors too faint to be seen with the naked eye, but that can be detected using telescopes, special video cameras, or (more usually) radars. These particles have masses between about a millionth and a hundredth of a gram. As a rough estimate, about a thousand billion (10^{12}) of these pepper the Earth daily. Because the surface area of our planet is about 500 million square kilometers, one of these enters the atmosphere above each 500 square meters each day, this being about the area taken up by a typical suburban house and garden. Each of these particles has energy equivalent to a small firecracker, so that if they made their way through the atmosphere unimpeded they would be forever causing cracks and bangs all over the place, and that is contravened by our personal experience: You are not personally being peppered by fast-moving meteoroids because the atmosphere does act as a shield of sorts. The effectiveness of that shield we discuss in Chapter 9.

The 40,000 tons of smaller particles arrive, then, in a non-random fashion, with meteor showers being evidence of this. The question that needs to be addressed is whether the larger masses, comprising about 160,000 tons per year, arrive randomly or non-randomly in time. For example, if the meteoroid streams producing annual meteor showers also contained large bodies—and we know that they contain at least their parent comet or asteroid, and such parents are known to break up, producing swarms of large fragments—then particular weeks of the year might be more dangerous than others. Many estimations of the impact hazard—in particular the relative risk posed by 1-kilometer asteroids as opposed to 100-meter objects—are unwittingly predicated upon an assumption that impacts of all dimensions are sporadic, occurring randomly in time. A vivid example contradicting this has recently been shown to us when the fragments of Periodic Comet Shoemaker-Levy 9 smacked into Jupiter—getting twenty or more impacts in a week does not fit in with the previous presumption of random impacts occurring about once per century. Even simple naked-eye observations of meteors that you could do from your backyard prove false the assumption of a random influx to the Earth.

This is a pretty bold statement. To determine whether it is correct, some knowledge of the ways in which meteoroids/asteroids/comets behave in space is necessary if the problem is to be properly assessed (discussion to follow). That many scientists have made erroneous assumptions in their work can be seen as follows. In Chapter 6 we saw that the terrestrial evidence indicates that craters, geological upheavals, and mass faunal extinctions follow a cycle of about 30 million years (although there is some debate concerning the exact periodicity). Although some craters undoubtedly are formed sporadically (that is, randomly in time), it seems that at least 40%, and probably more, are produced on a definite cycle, the origin of which is likely related to the motion of the solar system up and down through the plane of the Milky Way. Now, contrast that knowledge with the fact that comparisons between

- cratering rates for the Earth and Moon based on counting craters formed over geological time and
- theoretical cratering rates calculated on the basis of observed asteroids and comets

assume that one can equate the two. That is, the clearly fallacious assumption is made that craters have been produced at a more-or-less constant rate over the past 3.8 billion years; fallacious, because we are now fairly certain that terrestrial (and hence lunar) craters are, to a large extent, formed during distinct epochs of higher impactor flux. Up until now the apparent near-agreement between the two data sets has been used as a fundamental assumption in assessing the hazard to humankind. The assumption is specious, however.

The problem is one of paradigms. The existence of a pervasive paradigm is one of the worst problems that faces science in all its manifestations. The concept of a paradigm *shift* is often exalted by the very people who do most to obviate such changes in doctrine, their conservatism being the barrier. For the majority of the time, dogma rules supreme. I can give a good example of this from the very field covered in this book. The main belt of asteroids, between Mars and Jupiter, is perceived by most astronomers as being primordial—the debris left over when a major planet failed to accumulate due to Jupiter's gravitational stirring. Meteorites are believed to be chunks delivered into Earth-crossing orbits from the main belt. As time went by, with more distinctions between meteorites being identified and thus more classes defined and likewise more asteroid reflectivity/color categories distinguished, the original idea of a few massive parent bodies that broke up was expanded to require a dozen fragmented parents, and then two dozen, and so on.

Few scientists question the initial paradigm. Astronomers attend conferences each year bringing new, apparently startling discoveries that are meant to help us understand what is going on, but instead, confuse things still further. The answer, I believe, is with a heterodox view: The initial presumption is incorrect, and the main belt is not being depleted to supply meteorites and Earth-crossing asteroids, but quite the opposite. Reverse the arrow of time and the admirable (and complex) mechanisms discovered over the past decade or so, which explain how particles can be perturbed *out* of the main belt into eccentric orbits, can also result in objects leaving such eccentric paths and obtaining comparative stability in near-circular, main-belt orbits. The main belt would then be *gaining* mass, not losing it. The ultimate source of the fragments deposited in the belt also ceases to be a problem: Giant comets provide a ready supply, their differentiated interiors producing the necessary conditions for the variety of physical histories portrayed by our examination of meteorites—a paradigm shift waiting to happen, but don't hold your breath.

Now, back to meteor showers and their implications. Any object in orbit around the Sun does not remain in exactly the same orbit when time scales of at least centuries and millennia are considered. The gravitational tugs of the planets—either on other planets or on asteroids, comets, and meteoroids—cause the orientations of the orbits to swivel around. This swiveling is called *precession*; it is the same action displayed by a toy gyroscope or spinning top, although the toys precess much more quickly. Precession can best be visualized by thinking about the elliptical orbit of any object gravitationally bound to the Sun: That ellipse has a long axis (called the *major axis*), and as the object performs many circuits of the perimeter of the ellipse, the long axis slowly swivels around, taking thousands of years to perform a complete rotation through 360°.

Now, think about that ellipse in three-dimensional space. The ellipse is constrained to a plane, and in general, that plane is at an angle to the ecliptic (the plane of the Earth's orbit). This angle is called the *inclination* of the object's orbit. The ellipse will cut the ecliptic in two places: once as it passes down through the ecliptic plane and once as it rises again through it. These points are called the *descending* and *ascending* nodes, respectively. The Earth's orbit is also an ellipse but of low eccentricity, such that we could think of it as being close to a circle on the ecliptic plane. One or both of the object's nodal points may be close to that circle at any time, but as the orbit precesses, the points eventually will meet the circle, most likely at different times. In such epochs a collision between the Earth and the object—whether asteroid, comet, or meteoroid—is possible. What would be required for a collision is for the object to come to its node at exactly the same time the Earth is passing that point. That is not very likely—we have estimated the probability of its occurring earlier—but it does happen, as shown by impact craters. Indeed, the evidence is before your eyes every night as you watch shooting stars: Each has a very small chance of hitting the Earth's atmosphere, and yet you see a plentiful supply raining down.

Any individual object has its own elliptical orbit, but in the case of a meteoroid stream, what happens is that there are many meteoroids with very similar orbits forming a kind of tube. The orbits are similar because the meteoroids originated from the same progenitor (the parent comet), but the slightly different velocities with which they leave that parent lead to their dispersal into slightly different orbits. The typical width of such a tube can be gauged from my earlier statement (which you can easily verify) that meteor showers persist for about a week, with peak activity usually on just one or two days due to there being more meteoroids concentrated toward the center of the stream. In a single day, the Earth travels about two and a half million kilometers, so it is clear that meteoroid streams can be quite wide. If we return to the idea of the nodal points of an orbit cutting the ecliptic plane, for a meteoroid stream, the situation is better thought of as resulting in a large area rather than a point, the Earth passing through that area and the meteor shower continuing until we are clear of it. In fact, for most streams, that area is really an elongated oval rather than a circle (that is, the stream is not circular in cross-section), but that is a detail of no concern here.

As the stream precesses, that area (the region encompassing the nodes on the ecliptic of the individual stream meteoroids) will move inward and outward slowly so that in some epochs the Earth will be passing through it, while at others, the stream is not intersected by the Earth at all. How does this correlate with observations? The answer is, very well. For example, the Geminid shower appeared in the last century, not having been seen previously (at least since the Middle Ages), with a gradual increase in activity from year to year being observed, until currently it is one of the most intense and regular showers. Calculations based on its precession rate indicate, however, that within another couple of centuries, it will have turned such that Earth intersection no longer occurs. The shower will disappear, to recur in perhaps a thousand years' time.

Such a stream may meet the Earth while the meteoroids in their orbits are moving toward the Sun, thus striking the Earth on the nightside and producing a shower observable by the naked eye; the stream can also meet the Earth after passing perihelion, however, and then the meteoroids would arrive on the dayside and be observable only by radar. This would be at another time of year. In fact, the Geminid stream does produce another shower, identified by radar, called the Daytime Sextantids (because the shower radiant in this case is in the constellation Sextans).

Things can be even more complicated. Small differences between the sizes of the orbits of the meteoroids lead to differing precession rates and cause the stream to become dispersed—not dispersed such that the orbits are random, but instead such that their major axes have all orientations from 0° to 360°. This would imply that the area delineating their nodal points on the ecliptic not only becomes a slightly elongated oval, but in fact stretches all the way from its perihelion to its aphelion distance. Under these circumstances, the showers are seen continuously year after year (but still only for a limited season), not just for a few centuries; as well, during the showers, meteoroids are seen that pass both the ascending and the descending nodes. The best example of this is the Taurid showers, which have long-lived peaks in October-November, but with some shower meteors being seen all the way from September through to at least January. One shower (the Northern Taurids) has its radiant in Taurus but north of the ecliptic, due to meteoroids passing their descending node, while the other (the Southern Taurids) has a radiant south of the ecliptic, reflecting impacts at the ascending node. Those are the nighttime showers caused by meteoroids moving toward the Sun; there is also a pair of radar-detected daytime showers, due to meteoroids receding from the Sun, called the Beta Taurids and the Zeta Perseids. These are active in May-July, meaning that the Taurid Complex (which is discussed in more detail later) has four distinct, but related, meteor showers. I will not bore the reader by explaining how complicated orbital evolution phases can cause some streams to produce eight or more showers. We now have the fundamentals to understand the discussion presented here.

The Geminid and Taurid streams both precess continually, their major axes spinning around to take all values from 0° to 360° before cycling again. Not all streams do this, however. For example, Comet Halley and its stream nod backward and forward between two extreme angles, a motion termed *libration*. As a result, its two main showers do not become widely dispersed and will remain active for millennia because they will not precess so as to miss the Earth. One might also note that the comet itself currently has a nodal point far from the Earth's orbit, implying that (1) the comet cannot hit the Earth in the present epoch and (2) the meteoroids that we observe in its showers must have been released thousands of years ago when P/Halley had a node near our planet. Because of this, it was to be expected that there would *not* be enhanced activity from the Eta Aquarids and the Orionids in the 1980s while the comet passed through the inner solar system, unlike the case for the Perseids (because their parent comet—P/Swift-Tuttle—*does* have a node near the Earth, many of the meteors observed in the past few years having been released in the last perihelion passages in 1862 and 1737).

What happens when a meteoroid is released from a parent comet (or asteroid)? How does it move relative to the much larger parent? Comets are most active when they are closest to the Sun, because the higher solar radiation flux causes more of the volatiles to evaporate and thus more meteoroids to be released. Because the comet nucleus is spinning and tumbling, the meteoroids overall are released in all directions with more or less the same speed (the reasons for their ejection speeds will be indicated later). Sideways ejections, however, have little effect on the resultant orbits, most meteoroids moving predominantly along the same direction as the cometary motion. The major effect is that the sizes of these meteoroid orbits are increased or decreased, meaning that their orbital periods differ from that of the comet—those meteoroids ejected forward have larger orbits and so take longer to come back to perihelion, while (paradoxically) those ejected backward have lower orbital speeds, thus smaller orbits, and come back to perihelion sooner. Consider, for example, a comet with perihelion at 1 AU and aphelion at 4 AU: Its speed at perihelion is about 37.66 kilometers per second. A meteoroid ejected forward at 100 meters per second will return to perihelion 4.08 years later, while one ejected backward at the same speed returns to the same point in just 3.83 years. This means that it takes just 16 orbits (or 60 to 70 years) to form a complete loop, and because meteoroid streams have lifetimes of tens of thousands of years, it is not surprising that, in general, little annual variation in activity is seen—nor is it surprising that we see meteors from particular comets (such as P/Halley), despite the fact that the comet in question is far away. That is, we see the meteor showers associated with P/Halley every year, not just once every 76 years.

Is 100 meters per second a reasonable ejection speed, and how are such speeds attained? In fact, this has been a long-term problem in solar system astronomy. We have seen that meteor showers last for around a week, and from such durations we can calculate the width of the stream. That width will be due to a number of effects, governed by the *sideways* ejection speeds of the meteoroids. Such a width is indeed diagnostic of ejection speeds of 100 to 500 meters per second. In fact, it is found that meteor showers, the Geminids notable among them, last for different lengths of time, depending on the sizes of the meteoroids being studied. Thus the optically observed Geminids, produced by marble- to orange-sized meteoroids, last only for a few days, whereas the radar-detected Geminids, caused by sand- to pea-sized meteoroids, last for a week to ten days. These durations reflect ejection speeds of around 100 meters per second for the larger meteoroids and 500 meters per second for the smaller ones.

The question of how those speeds are attained has not proven to be an easy one to answer. Back in 1951, Fred Whipple published an important paper in which he considered meteoroids to be small solid objects released from the surface of a comet and swept outward from the cometary nucleus by the expanding gas produced by evaporating volatile chemicals in the nucleus. The relative velocities calculated in this way, however, turn out to be much too low to explain the duration of meteor showers, the Geminids in particular. Over the intervening decades, many astronomers, myself included, have scratched their heads and tried in vain to reconcile Whipple's model with actual observations. Perhaps we should have recognized earlier that there was a problem with the assumptions made, in that the model also leads to a prediction of some maximum mass meteoroid that can be ejected from a comet, because the gas outflow cannot overcome the gravitational pull of the comet on meteoroids larger than about a ton, and yet some very bright fireballs apparently in streams (and so apparently released by comets) have masses greater than this limit.

The solution to the conundrum has been provided by the spacecraft data returned from Comet Halley in 1986; having stated that, it must also be said that the details are not yet worked out. The various spacecraft and, in particular, the European Space Agency's *Giotto* satellite, which went in closest, showed that the coma did not decrease in density, as would be anticipated if all the gas were being *directly* produced by evaporation of volatiles from the solid nucleus. In fact, the source of the coma, at least in part, was the coma itself, in that solid lumps of material (that is, meteoroids) were leaving the nucleus and joining the coma. The volatile materials of which these were largely comprised were then exposed to sunlight and, of course, started to sublimate. The coma therefore had an extended source rather than a small point source (the nucleus): Remember that a cometary coma is tens of thousands of kilometers in size, whereas the nucleus is only a few kilometers across in most cases.

This was of interest to those space scientists investigating cometary comae, but it also has implications for afficionados of meteor showers. Comets suffer from nongravitational forces due to this outgassing and release of solid material during each perihelion passage, for example, P/Halley returning four days late in each apparition compared with what would be expected if only the gravitational pulls of the Sun and the planets were significant. That four-day difference is small compared to its 76-year orbital period, because those nongravitational forces are having to push a large body around. That is not the case, however, for a small meteoroid released by a comet: If one imagines a 10-gram meteoroid, perhaps made up of 80% volatile chemicals and 20% refractory material, it is easy to see that by the time the volatiles have sublimated, imposing a jet force as they do so, the remnant 2-gram meteoroid might be moving at an appreciable speed relative to its parent comet, having done a random walk in space to attain its final velocity vector. It would be similar to inflating a balloon and then releasing it. Who knows the direction in which it will end up after whizzing around the room?

This, then, seems to be the origin of the ejection speeds of meteoroids from comets. It also supplies an explanation for why larger meteoroids are ejected more slowly than small ones, as a consequence of

Newton's F = ma: the larger mass has a smaller acceleration in the same way that a truck has limited acceleration compared with a sports car, despite the truck having a more powerful engine. In view of this, we might expect smaller meteoroids to be distributed equably around the stream orbit, with larger ones preferably grouped near the parent comet. Again we ask whether this conjecture is confirmed by observations, and again the answer is yes.

To illustrate this point, about once a decade or so, meteor watchers are treated to a display of exceptional beauty when myriad streaks across the sky are seen in a super-intense shower. Such an event is called a *meteor storm*, with some displays producing more than 100,000 meteors an hour—the sky looks as if you are moving through a snowstorm. An example is the Leonid shower that occurs in its weak form each November. This shower is known to be associated with Comet Tempel-Tuttle, which has an orbital period of just over 33 years. Leonid storms tend to be seen when Comet Tempel-Tuttle is passing close by the Earth on its voyage around the Sun. Historically there have been many exceptional storms, for example, in 1799, 1833, and 1866. The realization in the nineteenth century that these meteors originate from P/Tempel-Tuttle was an important step in solar system astronomy. There was no pronounced storm in 1899 because the comet's period is not exactly 33 years. There was a less prominent (but nevertheless spectacular) storm in 1933, but in 1966 an awe-inspiring storm was witnessed. Planning is already underway for an anticipated humdinger in 1998 and/or 1999. Meteor enthusiasts are monitoring the annual Leonid shower to see whether it is gradually increasing in activity, so that they are ready for a major storm if it arrives. A few other comets also produce storms in the present epoch, for example, P/Giacobini-Zinner, which occasionally renders a storm (called the October Draconids) on a 13-year cycle, the last occurrence having been in 1985. That is not to say that all meteor storms can be predicted in advance, because occasionally one catches us totally by surprise, with no prior meteor activity having been noted and no parent object known—for example, the brief Phoenicid storm of 1956, which was seen from Australia to South Africa.

The deduction from the occurrences of the Leonid storm, and the "missing years," is that there must be a substantial concentration of meteoroids close to the nucleus of P/Tempel-Tuttle in a very restricted volume of space. Because the storms are much more prominent among visual meteors than radar-detected particles, it follows that larger lumps of material are congregated closer to the comet nucleus than the smaller ones. This is entirely as expected, because smaller particles will attain larger ejection speeds and so quickly move away from the nucleus. The bottom line is that we expect to find the largest fragments that have peeled off of any comet close by the nucleus and in an elongated core stretched along its orbital path but only filling a small portion of that orbit. A meteor storm will occur whenever, by chance, the Earth passes through that core.

Again, we must ask whether there is any direct evidence for the existence of these elongated cores, a question that can again be answered in the affirmative. One of the big surprises to come out of the Infra-Red Astronomy Satellite (IRAS) project, which was mentioned earlier in connection with the discovery of asteroid 3200 Phaethon, was the discovery of cometary *trails*. Note that the word is trails, not tails: Comet tails have been well known for millennia (the word comet meaning "hairy one") and are produced by the gas and fine dust released by a comet as it approaches the Sun. Comet trails are quite different, however, and were unexpected before their discovery in 1983. Whereas comet tails point radially away from the Sun, due to the solar wind and solar photon pressure driving the atoms/ions and tiny dust grains outward, a comet's trail really does trail behind. Trails have been found in association with a handful of short-period comets that were in appropriate positions to be scanned during the 11-month IRAS observation program. They showed up as bright infrared emitting bands stretching both forward and backward along the orbits of those comets but only occupying a restricted arc. It is the restricted length of that arc that results in meteor storms occurring only when the comet passes close by our planet, it also being necessary that the Earth's orbit intersect that of the comet to within the width of the trail, which is much narrower than it is long. It is therefore unsurprising that meteor storms do not occur more frequently.

Now is the time that we have to admit our ignorance. We really do not know how big the particles are that make up those trails. It might be that they are all smaller than the size of an orange or it could be that there are many 10- to100-meter objects in the trails. IRAS was a pioneering program that told us far more about the infrared universe than we knew previously, but it was a relatively unsophisticated instrument compared with three distinct infrared satellites that are due for launch in the next few years by various national consortia. It is hoped that these will tell us much more about comet trails, because (as we will see) these structures could be of immense import concerning the extraterrestrial hazard to humankind.

This chapter has entailed a tortuous discussion of how objects behave in space, but I ask your forebearance because we are coming to the punchline. We know that the old gas-outflow model for the ejection of meteoroids from cometary nuclei, which would limit the individual masses that could escape a nucleus, is incorrect (or at least not the most important acceleration agent), although that mechanism is undoubtedly important for small dust grains. Much larger fragments of cometary material may leave the nucleus due to fracturing of the surface exposing them to sunlight, their own jet forces then accelerating them away from the remnant nucleus. Even if scientific objections to that scenario are found, one can point out that we have observed many comets to split into two or more large pieces from time to time (for example, Comet Biela in the 1840s), and as I edit this chapter for the final time, newly discovered periodic comet Machholz 2 has been found to have many fragments. That is, it is entirely to be expected that some comets, perhaps the majority, will crumble into smaller fragments either in single catastrophic episodes (like P/Shoemaker-Levy 9 and the Kreutz group progenitor[1] as they passed too close to massive objects), in gentler splittings in interplanetary space (as did P/Machholz 2 and P/Biela and more than a dozen other comets in the past two centuries), or as a result of gradual erosion. In each case, we expect there to be a complex of debris formed along the comet orbit, as evidenced, for example, by meteor storms produced by pieces of the disintegrated P/Biela in late November 1872 and 1885.

Meteor storms, themselves, are not dangerous, except perhaps to astronauts and satellites, because small meteoroids burn up in the atmosphere. But much more infrequent than a meteor storm is an asteroid storm, when by mischance the Earth passes right through the orbit of a comet either just before or just after the comet comes through that point; recall that the larger fragments will be concentrated close to the nucleus. Such close transits may be rare, but they will occur much more often than comet nucleus impacts on the Earth. And if a comet has an associated complex of substantial (50- to 300-meter) objects co-orbiting with it, then clearly we would be in trouble. The sky would light up, either night or day, with a spectacular influx of shooting stars and fireballs, but with them would come rocks as large as a city block which would punch deep into the atmosphere before detonating, wreaking death and destruction on the inhabitants below.

For this apocalyptic idea, I coined the term *coherent catastrophism*, the cataclysm being spread over a wide area of the planet due to the coherent arrival of many impactors within a day or two, as opposed to the sporadic (or stochastic) impacts anticipated if objects are spread randomly in space. It is entirely feasible that within those few days the Earth would receive hundreds of blows like that of the Tunguska object.

The reader might find this to be too fantastic (in the literal sense of the word) to be credible and prefer the simplistic assessments of the hazard to humankind (the once-per-100,000-year one-kilometer asteriod impact)—that is, the idea that it is only the very occasional impact by a large asteroid or comet that is significant. If this is the reader's response, let me now point out that such assessments neglect the majority of the mass delivered into Earth-crossing orbits. How can that be? The answer is quite simple.

We have discussed the mass that strikes the Earth each year, with the year-in, year-out total mass of around 40,000 tons in the form of small (subgram) particles, which either ablate in the atmosphere as meteors or are so small that they gradually decelerate and then settle out to the surface. We also noted that this is by no means the total mass accumulated by the Earth, because the majority (estimated at around 160,000 tons per annum) comes in the form of massive bodies—the very infrequent asteroids and comets that have been discussed throughout this volume. The question is, "How big is the largest one?" The reason that this is important is that, using the observed size distribution of these bodies, most of the mass must be held in the few largest objects. For example, the largest known Earth-crossing asteroid is 8 kilometers in size, which means that it has a mass 500 times that of a 1-kilometer asteroid; thus eight or ten 8-kilometer (or larger) asteroids could have a total mass in excess of that of the few thousand 1-kilometer or so Earth-crossers. Similarly P/Halley and P/Swift-Tuttle, being 10 to 20 kilometers in size, dominate the total mass held in short- and intermediate-period comets.

But these are only the objects that we currently observe. In reality, there have been larger Earth-crossers in the past. There is clear evidence of this from both lunar and terrestrial craters. If the Chicxulub crater is in fact 400 kilometers across, the impactor may have been 20 kilometers or more in size, and even if the initial size estimate of 180 kilometers is accepted, this still implies an impactor larger than any known Earth-crosser observed in modern times (the last century or so). I was once a graduate student at the University of Colorado at Boulder, and I recall, maybe not verbatim, an inscription over the grand entrance to the Norlin Library there: "Who knows only his own generation remains forever a child". Similarly, it could only be a child who would believe that what we observe in the present epoch is diagnostic of the long-term conditions. This sort of view, which has gradually been pushed out of the geological fraternity if not the astronomical one, is broadly termed *uniformitarianism*; as time passes, it is becoming increasingly evident that this view does not represent the real universe.

What, then, is the most massive comet? I concentrate here on comets because it seems that the most prodigious bodies delivered into Earth-crossing orbits with regularity (on the astronomical time-scale) are indeed comets arriving from beyond the planetary region, rather than asteroids from the main belt. Our first approach to the problem might be to search through historical records. For example, in 1729, a brilliant new comet was discovered—Comet Sarabat. This was intrinsically the brightest comet observed in recent centuries, and a lower estimate of its size is about 100 kilometers; actually, it might have been up to 300 kilometers across. Although this comet had perihelion near 4 AU, and so could not hit the Earth, it is inevitable that many similar comets on Earth-crossing orbits have arrived over geological time. Indeed, the records of humankind's various civilizations in the past few millennia record several such sightings.

Of course, many will be dubious about accepting ancient observations,[2] and so we turn to modern data. In 1977, Charles Kowal, then working at the Palomar Observatory, discovered an object on an orbit crossing the planets Saturn and Uranus. It is thus intrinsically unstable because close approaches to those planets will inevitably perturb its path. This object is known as asteroid 2060 Chiron[3]; it is about 300 kilometers in size. Initial investigations by a number of people showed that its orbit was indeed only stable for a length of time much less than the age of the solar system. We cannot be certain exactly what its path will be or where it has been in the past, but in 1990 my colleagues Mark Bailey (of Liverpool John Moores University in England) and Gerhard Hahn (now with the German Space Research Agency in Berlin) conducted an extensive set of numerical integrations of Chiron using a large set of start orbits that were similar but which differed by small amounts within the bounds of our uncertainty in Chiron's position. Bailey and Hahn found that there was a very strong chance that within a million years Chiron will migrate into an orbit passing through the inner solar system. Such an event would spell disaster for humankind even if the Earth did not receive an impact by Chiron itself, or even any large lumps, because the amount of dust left in the atmosphere would lead to a significant cooling of our environment. Chiron's mass is many times higher than that of the cloud of material supplying us with 40,000 tons of dust a year, and that influx would be incremented by about two orders of magnitude should Chiron enter an Earth-crossing orbit.

In the 1980s, it was still possible for people to dismiss Chiron as being an oddball, a unique phenomenon that does not tell us much about what happens here in the inner solar system. But then, a whole series of discoveries in the early 1990s meant that one had to accept Chiron and its ilk as commonly occurring objects. In early 1992, another massive outer solar system object, 5145 Pholus, was found independently by the Spacewatch team at the University of Arizona and by Gene and Carolyn Shoemaker and David Levy at Palomar. The following year, Spacewatch found 1993 HA_2. That made three inherently unstable 100-kilometers-plus objects on orbits crossing the giant planets, any or all of which might plunge into an Earth-crossing orbit within a few million years. If there are three of such size detected, then it is safe to assume that there are dozens more awaiting discovery and, presumably, myriad smaller objects too faint to be detected with the telescopes presently being used in the searches.

Again, the disbeliever could claim that we only have theoretical evidence for the idea that these objects could enter the inner solar system, so a discovery made by my colleague Rob McNaught in early 1991 is important. He found an asteroid that is on an orbit totally unlike any other known to date. Chiron, Pholus, and 1993 HA_2 have low-eccentricity orbits in the outer solar system, but this new asteroid (5335 Damocles[4]) was found to have a high-eccentricity orbit passing from near the aphelion of Mars out as far as Uranus, taking 42 years to complete each lap around the Sun. Damocles, which is estimated to be 15 to 20 kilometers in size, therefore seems to represent an object in transition between a near-circular outer solar system orbit and an eccentric trajectory taking it through the inner solar system. Future projections of its dynamical evolution carried out by David Asher (now my colleague at the Anglo-Australian Observatory), Bailey, Hahn, and myself show that Damocles may well attain an Earth-crossing orbit within 100,000 years. Lest the reader think that result implies that we are safe, it should be noted that Damocles very nearly slipped through unnoticed. The probability of discovering such an object on each perihelion passage is less than 1%; that is, it is very likely that there are hundreds, possibly thousands, of objects like Damocles that have yet to be spotted (and even more of smaller size).

This chapter is full of questions. That is one of the reasons that planetary science in general, and the study of impacts specifically, is so exciting: We are still so ignorant of what is going on, and there is so much to learn. Earlier we mentioned that because Earth-crossing bodies have collisional lifetimes far less than the age of the solar system, a continuing source is required. The same idea applies to outer solar system bodies: If Chiron and its like have dynamical lifetimes of only a million years or so, they must be being supplied from somewhere. This is the next problem to need a solution.

One possibility is that Chiron and the myriad other outer solar system comets and asteroids are, in fact, captured Oort cloud objects. We have already discussed (in Chapter 6) the evidence for a comet wave from the Oort cloud having arrived in the recent past, meaning the past few million years. Thus perhaps, as indeed would be anticipated, some of those objects were captured into quasi-stable outer solar system orbits, and we are now observing the remnants that have survived until the present epoch. Certainly the expected timescales (millions of years) fit.

There is another possibility for which there is direct observational evidence. In 1951, the Dutch-American astronomer Gerard Kuiper hypothesized that there was no requirement for the planetary region to suddenly stop at Neptune and Pluto. He suggested that there might be a band of comets and planetesimals around 50 to 100 AU from the Sun. These would be the basic building blocks that had not managed to accumulate into a major planet. Kuiper's concept was for a flattened disk of objects occupying the same general plane as that of the planets; this contrasts with the Oort cloud, which had been suggested as being a spherical distribution of comets much further away and comprising the source of the observed long-period comets. Oort's idea was based on observational evidence (the distribution of the energies and sizes of long-period comet orbits), whereas Kuiper's was a theoretical construct, now known as the Kuiper belt. There was no real evidence for its existence until late in 1992, when David Jewitt (University of Hawaii) and Jane Luu (now with Harvard University), after many years of trying, identified an object about 200 kilometers in size beyond Neptune. This was by far the most distant asteroid or comet (whichever it may be) to be identified by that time.

As is so often the case, things are easier when you know *how*. In the two years following the discovery of the first object, another 16 trans-Neptunian bodies were identified. All are larger than 100 kilometers in size, and, as usual, it is to be anticipated that they are accompanied by a much greater number of smaller objects that remain beyond the grasp of our telescopes. I would be surprised if this total of 17 bodies has not been incremented before this book is published.

Due to the fact that at such a distance objects move very slowly about the Sun, and also that astrometric observations have been sparse, it is not yet clear whether these bodies are true Kuiper belt objects or perhaps Neptune Trojans[5]; the odds are on the former, however. If this is the case, then again they will be dynamically unstable and will eventually either be ejected from the solar system or passed on inward from Neptune to the other giant planets. In a small fraction of cases an Earth-crossing orbit will be attained.

There is no real difficulty, then, in believing that giant comets must arrive in Earth-crossing orbits from time to time; indeed, the difficulty would be in explaining why they would *not*, given the data in hand. To reiterate, the majority of the mass delivered to Earth-crossing orbits would be held in such giant comets, as shown in Figure 6. Should this fact be considered when making a hazard assessment?

The crux of this question is whether a giant comet has arrived *recently*. Bailey and colleagues have looked at how often one might expect giant comets to arrive in the inner solar system and be trapped in short-period Earth-crossing orbits. They found that the likely answer is about once every 100,000 years. This means that there is only a one in a thousand chance of a giant comet doing this within the next century, which means that from the perspective of the hazard to humankind, we might decide to note the possibility but not take any action; in our present technological state, it will likely remain in the "too-hard basket" for some time. However, if a giant comet has arrived within the last 100,000 years, as would seem likely from the odds, then this would have important implications for humankind.

Now, obviously there is, at present, no giant comet trapped in a short-period orbit with the exceptionally bright comets observed in historical times having predominantly been on parabolic orbits such that they will not be seen again by our civilization. But any giant comet that by chance has been captured, most likely by Jupiter, into an orbit within the inner solar system would not be expected to remain intact for tens of thousands of years. We have discussed the numerous reasons why comets break apart (for example, near-passages by a planet [especially Jupiter] or the Sun, thermal stresses upon a weak structure, meteoroid impacts, and so on). Even in remote space, far from any planet, comets have been observed to split apart into multiple fragments, with timescales of only centuries or millennia being appropriate. Such a disintegration would be expected to continue until such time as the various fragments have sufficient cohesive strength to be able to survive.[6]

We therefore ask what we might expect to observe now, should a giant comet have broken up on some sort of orbit crossing that of the Earth and also, presumably, Jupiter (since jovian perturbations would most likely be responsible for the capture from a parabolic orbit). The first thing we might note is that it is known that such a capture is more likely to occur from a low-inclination orbit and the daughter orbit produced is also most likely to be of small inclination to the ecliptic. Let us start with the very smallest particles and then move upward in size.

The smallest particles we observe are those in the zodiacal dust cloud. These are indeed of low inclination. More importantly, however, it has been known for some years that the population observed now does not appear to be in balance with the supply from the comminution of meteoroids in catastrophic collisions: There are too many meteoroids and too little dust. That is, there is evidence that some exceptional event has occurred at some time within the survival lifetime of the meteoroids and dust, that lifetime being between 10,000 and 100,000 years.

Moving up in size from dust, we come to those meteoroids, in particular those around a millimeter in size which are radar-detected. Earlier in this chapter, we discussed how the meteor count rate varies during the day and seasonally: The peak would be expected to occur in the summer due to the relative orientation of one's latitude and the point on the Earth that is the furthest forward in its orbit (the apex). Indeed, a broadly sinusoidal variation in count rate is found, but it would be anticipated that if meteoroids were randomly distributed around the Earth's orbit, then these sinusoids would be six months out of phase if one compared measurements made in the northern and southern hemispheres. Back in the early 1960s, separate radar meteor projects were carried out from Ottawa in Canada by Peter Millman and Bruce McIntosh and from Christchurch in New Zealand by Cliff Ellyett and Colin Keay. Millions of meteors were detected from each site, allowing the hourly count rates throughout the year to be determined accurately. What was surprising was that, when the results were compared, it was found that the seasonal curves were actually five months, rather than six, out of phase. This demonstrated that meteoroids are *not* equably spread around the Earth's orbit. The interpretation of this was not simple, and it was some years before the implications could be recognized, because back then we did not know enough about the lifetimes of particles in space to realize that it implied that there had been some enormous injection of meteoroids within the past few tens of thousands of years; that is, on the astronomical scale, in the very recent past.

Through the 1960s and into the 1970s, there were various sets of radar equipment around the world being used for determining the orbits of meteors. There also were a few optical set-ups for making photographic orbit determinations. The late Jan Stohl of the Slovak Academy of Sciences spent some time in Ottawa studying their meteor count rates and recognized that a possible interpretation was that there was a vast diffuse stream of meteoroids that had hitherto been interpreted as being sporadic. Using the radar and photographic orbits, he tried to sort out those meteors that could be part of this stream from those that really were sporadic. He found that at least 50% of the so-called sporadic meteoroids were in fact part of a high-eccentricity, low-inclination stream with aphelion close to 4 AU. It appeared that the Taurid stream, mentioned earlier, comprises one part of this broad stream. My own contribution to this work was that I had shown that meteoroids on orbits with aphelia near Jupiter would be thrown out of streams on a timescale shorter than their lifetime against being lost through other mechanisms (such as spiraling into the Sun or hitting zodiacal dust grains). This implied that any part of Stohl's stream that had started out with a larger aphelion distance either would have been ejected from the solar system by Jupiter or would have had their orbits semi-randomized; that is, a large part of the *other* 50% of the sporadic meteors could also be understood as originating from this stream. Only those that had quickly attained aphelia at 4 AU or less (through a form of radiative drag force or decelerative impacts with dust near perihelion) would be safe from major perturbations by Jupiter, and even then, precession and slow orbital evolution would disperse their orbits over a few tens of millennia such that only a broad diffuse stream would survive. These would be difficult to identify as long-lived showers unlike the short, sharp showers produced by more compact streams.[7] It is interesting that the same sort of consideration applies to comets as well as meteoroids. In 1994 Eduard Pittich (of the Slovak Academy of Sciences, Bratislava) and Hans Rickman (of the University of Uppsala, Sweden) showed that because short-period orbits are rapidly randomized by Jupiter, it is feasible that the entire presently observed short-period comet population was produced quite recently by the break-up near Jupiter of a giant comet—in fact, just the sort of scenario now being discussed.

There is evidence, then, from the zodiacal dust and from meteoroids of a recent, exceptional event in the inner solar system. The problem of the origin of the short-period comets may also be answered in this way. What of the Earth-crossing asteroids? The assumption of stochastic catastrophism with random impacts, as implicitly assumed in the Spaceguard report (see Chapter 11), would imply that such asteroids will be randomly distributed in their orientations around the ecliptic. This appears to be the case for the "typical" Apollo, with moderate eccentricity and aphelion in the asteroid belt. These may well be random objects that have been thrown out of that belt. Quite a different result is obtained, however, if one sorts out the Apollos that have similar orbits to the Taurid meteoroid stream. If one does that sorting purely on the basis of the semi-major axis, eccentricity and inclination (a,e,i) of the asteroids, there is no obvious reason *a priori* to expect these to be other than randomly oriented. In fact, when David Asher, Victor Clube, and I did this sorting, we found that about 20 of the 140 or so known Apollos have similar (a,e,i) to the Taurid meteors, and, of those, about a dozen are aligned with the Taurids. One can test for the probability of the alignment occurring by pure chance: The answer is far less than 1%. Our colleague Bill Napier did a different form of test and came up with similar results—most unlikely to have occurred by chance. Thus it seems that we have shown that the Taurid meteoroid stream contained asteroids, too, ranging in size from 10 meters to more than a kilometer.

In Chapter 2 we discussed evidence for comets that have become extinct or dormant and also pointed out that Comet Encke apparently "woke up" and became an active comet just over two centuries ago. In fact, it is more than half a century since Fred Whipple pointed out the relationship between P/Encke and the Taurid stream. P/Encke is about 5 kilometers across and larger than any of the asteroids found in the stream. It may, therefore, be correct to think of it as being the parent of the stream. On the other hand, there may well be one or more dormant comets in the stream that we have yet to identify and that may exceed P/Encke in size. If the Spaceguard program goes ahead, we might expect to find them. Otherwise, it may be a few centuries or more before they wake up and proclaim their presence in an obvious way. In the meantime, however, we can consider P/Encke as only the likely major fragment left from a much bigger object that has broken up, producing all of those asteroids and other debris. What can its behavior tell us now?

When Whipple was analyzing the Taurid showers, his masterstroke was that he recognized that although the meteor showers consist of meteoroids having inclinations of between 2° and 6°, and the comet has an inclination (i) close to 12°, this does not mean that they are unrelated. In this chapter, we have already exhaustively reviewed the way in which the nodes of an orbit move inward and outward on the ecliptic plane, but this was all based on the assumption that the only thing that happens is that the orbit precesses, meaning that it is only a case of the major axis of the orbit swiveling around, with no other orbital changes occurring. In reality, things may be a little more complicated, as in the case of P/Encke. It turns out that although it currently has $i = 12°$, this is only while it has nodes far from the Earth's orbit, whereas at such times as it, or its stream, has a node near 1 AU (making an impact possible and making its spawned meteoroids observable as meteors), it has a rather lower inclination. Whipple realized this fact on the basis of the analytical techniques that he had available in the 1930s and 1940s. Using modern numerical integrations, we have been able to show that he was broadly correct, subject to the assumptions that he needed to make.

The bottom line is that in the present epoch, the Earth is touched only by the periphery of the convoluted Taurid stream, meaning that the meteor showers that we observe now are weak compared with what must light up the sky when the Jupiter-driven orbital evolution has brought the stream around to allow the Earth to pass through its core. Even at present, the daytime Beta Taurid stream, when it reaches its peak in the last week of June, is one of the most intense of all of the annual meteor showers. But this is just the tiny meteoroids that burn up high above our heads. The problem is that the core is also the place where the large meteoroids and small asteroids will be concentrated, so that the sort of havoc visualized in the coherent catastrophism concept would be wreaked when we pass through the core.

That the complex of material associated with the Taurids *has* a core is known from IRAS observations, which show a trail apparently associated with P/Encke. This is not to say that all of the large lumps will be close by the comet or that trail. For example, the asteroids we have identified seem to be equably distributed around the orbit, and although we are currently passing through the periphery, there have been some substantial impacts in recent times. For example, as discussed in Chapter 9, the Tunguska object appears likely to have been a member of the complex, arriving as it did during the peak of the Beta Taurid shower, with an observed radiant consistent with this shower. Similarly, the seismometers operated on the Moon for much of the decade after the Apollo lunar landings registered, at the end of June 1975, their major episode of impacts on the lunar surface by meteoroids of up to about a ton. The astronauts left seismometers only on the nearside of the Moon, and impulses were registered only when the Beta Taurid shower radiant was exposed to that hemisphere. The high near-Earth flux of objects detected during that week (or so) was also identified through ionospheric disturbances detected over a wide part of the globe, assumedly reflecting increased ionization from the ablation of small meteoroids; unfortunately, it seems that no specialized meteor radars were operating at the time.

During the twentieth century, there have been a number of other anomalous phenomena observed either in late June or in October-November when the nighttime showers occur. These have been catalogued by my colleagues David Asher and Victor Clube, who believe that their occurrence-years follow a definite cycle, which is indicative of the orbital period of the core of the complex. Late in the last century, on June 25, 1890, a meteorite fell near Farmington, Kansas; this happens to be, by far, the youngest meteorite known (in terms of space exposure age), having been on a free orbit for, at most, 25,000 years after release from its parent, less than one tenth the exposure age of the next youngest meteorite. Its observed radiant and date of fall are both indicative of membership with the Taurid Complex, and this age is in agreement with the formation time of the complex, as derived from dynamical studies (discussed later). Going back further in time, American astronomer-geologist Jack Hartung has suggested that an impact on the Moon was witnessed by five men (and presumably others elsewhere) in Canterbury, England, in A.D. 1178, and recorded by the monk Gervase. Hartung has linked that apparent impact with the formation of the Giordano Bruno Crater on the Moon and believes that the projectile was part of the Taurid Complex, based on the date of the sightings in the third week of June. Although Giordano Bruno is known to be one of the youngest lunar craters, others have pointed out that the probability is very small that it could have been excavated so recently. This objection is based on an assumption that impacts occur randomly in time; if we are currently (currently meaning the last 20,000 years) in an epoch of much-enhanced impact rates, then the objection disappears.

The Taurid core is what should worry us, rather than the peripheral debris that seems to have struck the Earth (and Moon) in this century. If the hypothesis that the IRAS dust trail represents a core containing many large projectiles is correct, then the following is implied:

1. A large fraction of the objects on Earth-crossing orbits, of all dimensions, are the daughter products from the break-up of a giant comet some time during the past 100,000 years, dynamical

studies suggesting around 20,000 years as likely. All that is suggested here is a break-up similar to that undergone by P/Shoemaker-Levy 9 in 1992, except by a comet at least 100 kilometers across and in an orbit crossing from Jupiter to the Earth.

2. The core of the complex, in a coherent subjovian orbit, evolves to have a node near 1 AU every millennium or so, at which time the Earth is bombarded by many macroscopic objects in episodes at certain times of year. It is these events that dominate the hazard to humankind. Such an episode would last for a century or two.

If this is the case, then episodes like that mentioned in item 2 should be recorded in human history. Some of the evidence that such records exist is discussed in the next chapter. There is also evidence from modern-day investigations that shows, for example, that the amount of interplanetary dust may have been ten times higher 20,000 years ago than it is today (using an analysis of microcraters on lunar rocks), and that there was a much higher dust influx to the Earth's atmosphere 16,000 to 20,000 years ago (from arctic ice core analyses). Again, this evidence is discussed in the next chapter.

One final thought about the importance of meteors: If we are serious about investigating the chances that some large asteroid or comet is going to strike the Earth soon, with cataclysmic consequences, doesn't it make sense that we try to learn some lessons from the vast number of smaller objects that are striking the Earth now? This obvious idea has been overlooked by almost all of those, self-appointed or otherwise, that have considered the hazard posed to humankind by asteroids and comets. Our data resource on Earth-crossing asteroids and comets numbers fewer than 1,000 orbits, whereas we have measured more than 400,000 meteor orbits. Should we not, at least, take a look and see what those meteors might be able to tell us?

8

Stonehenge and the Pyramids

If one were asked to name the two most famous archaeological relics in the world, one could not go far beyond Stonehenge and the Egyptian pyramids. Certainly the Parthenon in Athens could stake a claim, as could the mysterious statues of Easter Island, the deserted cities of the Mayan civilization, Machu Pichu in Peru, and Mesa Verde in Colorado, or perhaps the Roman amphitheater known as the Coliseum. But Stonehenge and the pyramids seem to be special, perhaps because we recognize the huge amount of effort required for their construction, coupled with the mystique surrounding their purposes. We know that the pyramids served as tombs for the Egyptian kings and queens, but why such gigantic mausoleums, taking years, perhaps decades, to build and at extraordinary cost? And regarding Stonehenge, numerous suggestions have been proposed, but there is still no definitive explanation of its purpose. Could it be possible that the construction of these mysterious structures was prompted by calamitous events in the skies? It may just be that these archeological enigmas are further evidence of the history of the intrusion of rogue asteroids and comets into the Earth's vicinity.

In this chapter, I take the liberty of making that very suggestion about these ancient megaliths, a suggestion that many will find bizarre. Indeed, I cannot say that I more than half-believe my speculations in this chapter myself. However, playing the devil's advocate is a good way to get an argument going (as is my intention), and it is only if matters are exhaustively discussed that we will arrive at the truth. Thus I ask that you take this chapter with a pinch of salt, by which I mean you should take my arguments as a series of suggestions whose plausibility may be considered and then accepted or challenged.

The first thing that we should note is that astronomy, surveying, and timekeeping have been linked for utilitarian reasons ever since humans first built houses, meeting places, and constructions for ritual purposes (for example, churches) and began to keep a calendar. Anyone doubting this might like to consider how those of the Islamic faith align themselves with Mecca while praying, how Christian churches are oriented, or how the date of Easter is determined each year. It has only been in the past decade or two that laser theodolites, satellite navigation systems, and atomic clocks have replaced celestial observations as the best systems for fixing position and time. Visit a library that utilizes the Dewey Decimal classification system, find the number 520 (which is where astronomy starts), and discover which subjects follow on the shelves. The grouping you'll find is more than an anachronism; historically speaking, advances in time keeping and surveying have been tied to improvements in astronomical knowledge. Thus we might anticipate that ancient artifacts apparently aligned with celestial phenomena at certain times of year did not acquire their orientations by accident. That is, it would be reasonable to adopt the null hypothesis that the alignments are by chance, in the expectation that the hypothesis is disprovable at a high confidence level, and therefore the constructions understood to have been carefully planned by the ancients, who must have been cognizant of what was happening in the sky.

I was born not many miles to the west of Stonehenge and was often taken there as a child, when one could still walk through the circles and actually touch the stones. I have been back many times since. If one asks another casual visitor to that megalithic monument to define its purpose, the answer will generally be along the lines of, "I think it has something to do with astronomy and observing the Sun." What I am going to argue is that the first part of that answer is correct—that it does have something to do with astronomy—but that the original construction at Stonehenge was for observing an astronomical phenomenon that as yet has not been recognized. The conundrum of the original motivation of Stonehenge is abstruse and recondite, even profound, and has occupied archaeologists and astronomers alike for decades. I believe that the solution that I propose here has not been posed before, even though many disparate suggestions have been put forward, perhaps even as many as the multifarious conspiracy theories for the assassination of JFK.

We must first understand that the collection of structures that we call Stonehenge is in fact a series of developments that were built over a period of about 2,000 years, from around 3100 B.C. to 1100 B.C. It may disappoint you to learn that I am not concerned here with the major stone structures, the bluestones and sarcens, that are so famous. These central circular and horseshoe-shaped erections, only about 25 meters in diameter, comprise what is known as Stonehenge III and came much later than Stonehenge I, which today is easily missed by the visitor or is thought to be of minor importance. Stonehenge II, which was intermediary between the earlier and later developments, has few visible signs. See Figure 13 for a map of the various constructions, which will undoubtedly help with the following discussions.

Stonehenge I comprises a circular ditch and a bank having a diameter of almost 100 meters, just inside of which are 56 filled-in pits known as the Aubrey Holes. These holes were each one to two meters across and half to one meter deep. In addition, Stonehenge I includes a large rock known as the Heel Stone, which is near, but not on, the axis of symmetry of the whole site, pointing along a line running from near the northeast to the southwest; in fact, the azimuth of that line is just less than 50°. The fact that the Heel Stone is not *on* that axis led many to believe that Stonehenge I had a totally different axis to Stonehenge II and III, but in 1979, a refilled hole was discovered just over the axis from the Heel Stone, that hole bearing the impression in the subsoil chalk layer of the base of a stone that has since disappeared. Thus it seems that there were a pair of stones erected in Stonehenge I which straddled the axis.

There is another feature of Stonehenge I that must be mentioned.

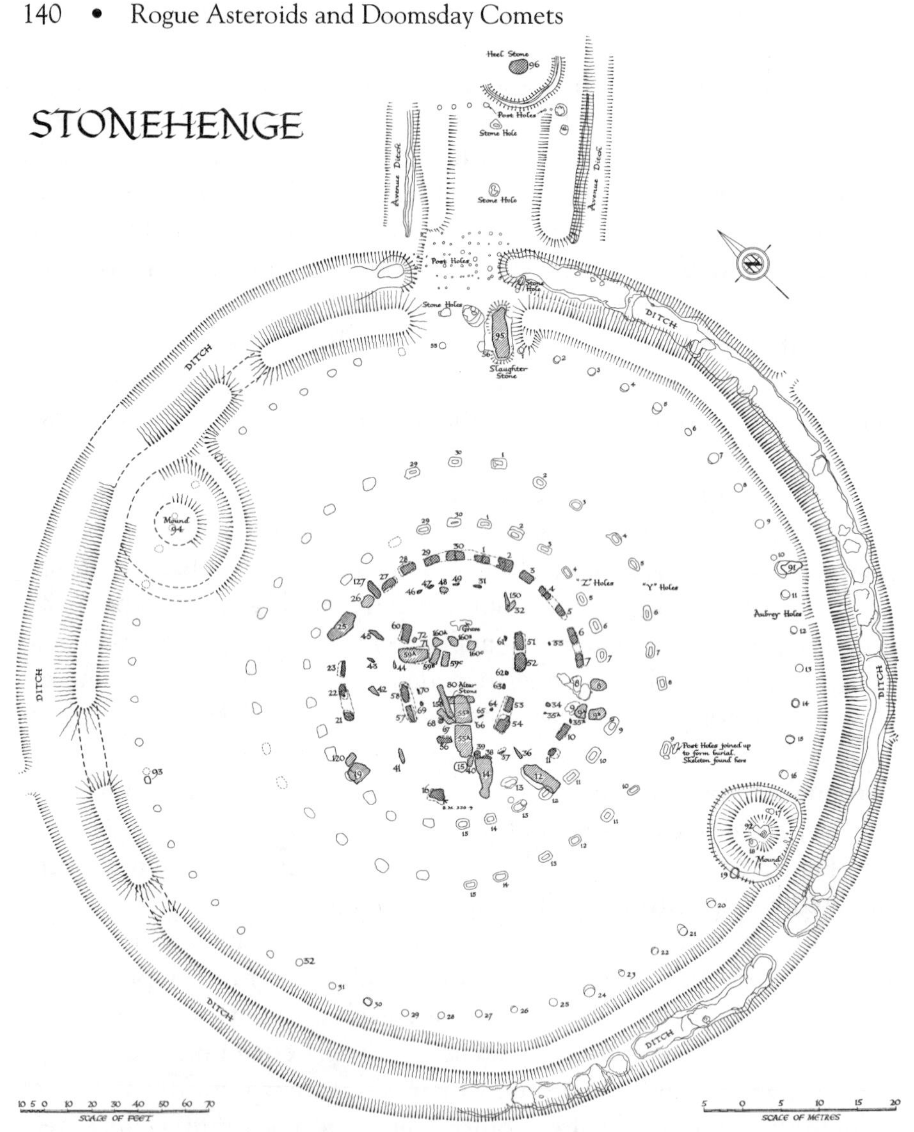

Figure 13. The ground plan of Stonehenge. Note that the original structure (Stonehenge I) comprises *only* the outermost bank and ditch, the 56 Aubrey Holes just interior to that bank, the Heel Stone and its absent twin, and *possibly* the four Station positions (91, 92, 93, and 94). (Courtesy of the Controller of Her Britannic Majesty's Stationery Office. British Crown Copyright.)

Overlapping the Aubrey Holes are four distinct features. Two are well-developed earthworks or barrows (one of which was first excavated in 1978) which have no associated stones, while the other two positions are defined by remnant stones; it may well be that at one stage there were stones at all four positions (which are called the Stations), but that two were carted off later or re-used in some way. The four Stations form a rectangle, with its long axis perpendicular to the main axis of Stonehenge. As will be seen, these four Stations are of great significance with regard to the astronomical interpretation of Stonehenge. The archaeological evidence puts the final development of these Stations later than the initial Stonehenge I construction, because the two barrows cut through and disturb the circle of Aubrey Holes. The conventional chronology has the Stations appearing as part of the initial stages of Stonehenge III, or at the very earliest in Stonehenge II, many centuries after the bank, ditch, and Aubrey Stones were dug out and the Heel Stone and its twin were erected. However, although the final form of the Stations, with elaborate earthworks and stones, might have come later in Stonehenge II or III, it is possible that the *locations* of the four Stations were used in Stonehenge I, if one accepts that Stonehenge is not randomly sited, geographically speaking. The reason for this is that at any other latitude a parallelogram rather than a rectangle would be needed in order to achieve the sight lines between the Stations which are aligned with certain positions of the Moon (discussion follows). This would imply that the site of Stonehenge I was carefully chosen and points toward Stonehenge's having been a luni-solar observatory *ab initio*. At this stage, the reader is asked to note that I have deliberately stated the viewpoint of someone with a set astronomical interpretation of Stonehenge (the Stations assumed to have been used by the constructors of Stonehenge I, which therefore fixes its latitude) rather than leaning toward the archaeological evidence for the Stations' having a much later provenance. The former view I will contradict in favor of the latter in what follows, this leading to an alternate astronomical hypothesis for the *origin* of Stonehenge, as opposed to its later *development*. Stonehenge I is the construction that we will concentrate on, after a brief review of the likely astronomical purposes of the remainder of the monument.

That Stonehenge has an axis of symmetry generally directed toward the point where the Sun rises in the northeast around Midsummer Day (the summer solstice) has been recognized for many years, with British astronomer Sir Norman Lockyer having done much to publicize this fact early in this century. The idea of Stonehenge having something to do with the Sun and astronomy has therefore had a long time to pervade the public thinking. The opposite direction, toward the southwest, is where the Sun sets on Midwinter Day (the winter solstice), and some have suggested that this, rather than midsummer, was the important time of year to the culture(s) that designed and built the monument.

In fact, the Sun does *not* rise at the same point every year; although the movement of that point is very small during any individual's lifetime. Over centuries, the change is noticeable. The reason for this motion is that the tilt between the Earth's spin axis and the ecliptic, an angle known as the *obliquity of the ecliptic*, slowly changes under the gravitational influence of the Moon and the Sun, with the other planets also having a minor effect. Lockyer used this fact to make an estimate of the age of Stonehenge, by calculating when the Sun would have been rising at midsummer on the projected axis of Stonehenge. Although there are many objections to the method he used,[1] Lockyer did derive an age (toward the middle of the second millennium B.C.) that is in reasonable accord with later archaeological studies of Stonehenge III, that utilized carbon-dating and other methods.

From the perspective of the astronomical interpretation of Stonehenge, the next great step forward did not come until about 30 years ago. As we have seen, the main axis marks the far northerly azimuth attained by the Sun at midsummer and the far southerly point in the winter. In 1963, a British Stonehenge enthusiast and amateur astronomer, C.A. (Peter) Newham, had an article published in a newspaper about his interpretation of Stonehenge, in which he pointed out that the sight lines between the Stations appear to be aligned with the maximum and minimum azimuths obtainable by the Moon each month (actually, each *lunation*). Variations in the monthly azimuths occur due to changes in the relative orientation of the pole of the lunar orbit to the Earth's spin axis. These follow a cycle of 18.61 years, this being the period required for the regression of the nodes of the Moon's orbit on the ecliptic. It had long been known that the Full Moon would appear once every 19 years over the Heel Stone at the time of the midwinter solstice in what is called the Metonic Cycle, and it had been suggested that the 19 bluestones in the central part of Stonehenge III were related to this number.

It appeared, then, that Stonehenge was both a lunar and a solar observatory, and it was a straightforward, if difficult, step to then show that it could be used as an eclipse predictor. This subject was taken up with gusto by Gerald Hawkins, an astronomer originally from England but who had been working in the United States for some years. Hawkins had been thinking about the Stonehenge conundrum at the same time as Newham, neither knowing of the other's progress. He showed that there were a whole series of apparent alignments of the stones of the Sarcen Circle, which he claimed to be associated with astronomical events involving the Moon and the Sun, with a very small probability that these could have occurred purely by chance. With such an interpretation, which has been argued against by many archeologists, Stonehenge may be viewed as a huge astronomical observatory; perhaps *observatory* is the wrong word; *predictor* might be more appropriate.

I have glossed over the intricacies of the work of Newham and Hawkins here, because they are covered in detail in many excellent books; for example, see that by Fred Hoyle[2] mentioned in the bibliography. To be fair to the archeologists, there were some dubious assumptions made by Hawkins which have been highlighted by Hoyle; this was also done in the book by Lancaster Brown. There has been more recent and important work by the late Alexander Thom (a Scottish engineer who had for many years been a professor at Oxford), along with co-workers including his sons, but since that has been well reviewed elsewhere (for example by Lancaster Brown, again) and does not impinge on the story in hand, I do not include it.

At any rate, the essence is that Stonehenge III (Stonehenge II was a brief, largely inconsequential episode) seems to have been built as a rather sophisticated astronomical device, but the positioning of the huge stones was predetermined, the builders knowing what they wanted to achieve. Since Stonehenge III was constructed a millennium or more after the first work at the site, it was built by an essentially distinct culture: How much do we have in common with the protagonists of the Battle of Hastings in A.D. 1066, or even the Pilgrim Fathers? Occam's Razor dictates that we look for an astronomical purpose behind the comparatively inelaborate design of Stonehenge I, which was built somewhere between 3200 and 2800 B.C. (and quite likely early in that period); this purpose was not necessarily the same as that behind Stonehenge III. The former was likely a true observatory (that is, a device for investigating unknown phenomena), rather than merely a predictor of anticipated phenomena as was Stonehenge III. What could that purpose be? What were the phenomena that so motivated the ancients?

I would like to quote from Hoyle's book, because he makes my point tellingly:

> Stonehenge I is essentially very simple, a set of marked positions and a few naturally occurring boulders—there may even have been wooden posts instead of boulders in the beginning. This simple structure was sufficient for the astronomical needs. Once simplicity became replaced by complexity, as in Stonehenge III, one can be virtually certain that science had been displaced by ritual.[3]
>
> In seeking the forerunners of Stonehenge I, we must look for something simple, we must look around the year 3000 B.C., not around 2000 B.C., and we must look in the right place, which is where?

Ah, hah! Now we're getting somewhere. Hoyle goes on to use an analogy from nature. An electric eel, he says, could not have evolved its ability to kill prey with an electric shock as a feature that always had that usage, because when it first appeared, the power available would have been insufficient to stun any potential meal. It seems that the eel developed the feature initially as a sensory organ of some type, and later evolution developed it into an ever more powerful weapon to fulfill a totally different purpose.[4] What does this mean for Stonehenge? According to Hoyle,

> In a like manner, eclipse prediction is far too developed and complex a concept to have existed in the beginning. The early astronomical ideas must have been used to solve less difficult problems. Just as the archeologist will not be satisfied until forerunners of the actual structure of Stonehenge have been traced, so it is necessary to trace the forerunners of the astronomical ideas themselves. It must have been possible to go from simple beginnings in a sequence of mounting complexity until the ideas necessary for eclipse prediction came at last into the perceptions of the early watchers of the sky.

Hoyle goes on to describe the astronomical predictions that the builders of Stonehenge III were seemingly able to accomplish when the great stone circles and horseshoe shape were constructed after 2000 B.C., but notes that due to a fluke agreement between certain periods, it all soon became rather routine—good for ritual, but too comfortable for scientific progress:

> But the 19-year eclipse-year cycle was a seriously regressive step. It worked—it was capable of working magnificently—but it led nowhere, because it was based on a mere fluke coincidence among the periodicities of the Sun, Moon, and nodes. It destroyed the need for meticulous observation. Stonehenge I would have worked just as well even if the fluke had not existed. But Stonehenge I and its concepts and ideas were gone by the second millennium B.C. The concept of constructing an instrument to observe the world was gone, and in that, much was lost.

That is, as stated earlier, Stonehenge III was not an observatory, but instead, more like an orrery. The only true astronomical observatory ever built at Stonehenge is the rather unimpressive Stonehenge I, all that remains of which is a smoothed-out bank and ditch, along with 56 refilled holes in the ground that are now difficult to identify, and a few other mounds and unshapen stones. Hoyle finishes his book by writing "In Stonehenge I may well be many roots of our present-day culture." This being the case, we would do well to identify the motivation for its construction.

As foreshadowed, we now return to the question of the Stations and their implications. In the vicinity of the earthworks that make up Stonehenge I, only the four Stations have a known, or plausible, association with the movement of the Moon. That is, although a strong case exists for the later development of Stonehenge as a luni-solar eclipse predictor, there is nothing to suggest that such a purpose was the aim of the initial builders *unless* one accepts that the Stations were in use by the constructors of Stonehenge I. The archeological evidence is that the final form of the Stations, with their attendant stones, came much later, although, as we have said, that does not preclude the locations of the Stations having been used at the time of Stonehenge I. Counting *for* the Stations being an entirely later addition, like the sarcens and bluestones, is the fact that they do not form any coherent link with the positions of the Aubrey Holes; counting *against* the Stations having been post–Stonehenge I is the apparent fact that their rectangular arrangement defined the place where Stonehenge had to be built. I want to consider the latter idea and give it an entirely new twist.

It has been stated herein that the Stations form a rectangle. Is that actually true? People have claimed that they form a perfect rectangle, implying that the four angles are precisely 90°, but as anyone who has conducted any scientific experiment knows, any measurement is subject to tolerances or uncertainties. Just how close are the Stations to being at the corners of a rectangle? Measurements made in 1978 by R.J.C. Atkinson, a British archaeologist who has studied Stonehenge for several decades, indicate that while one angle is very close to 90°, two others are around 89.5° and the fourth is close to 91°. Taking these at face value (that is, assuming that Atkinson provided reasonably precise measurements; he quoted the angles to one part in 600 of a degree), the deviations from a right angle may be expressed in terms of how far from the "correct" latitude Stonehenge could have actually been placed. That is, if the ancients who planned out the stations could only get the angles correct to within one degree, how far out could they have been in the latitude? The answer is about 40 kilometers. That does not sound like very much, but on the scale of the south of England it means that Stonehenge could equally well have been located on the south coast where the town of Bournemouth now stands, or as far north as Swindon. In fact, a 80-kilometer wide band covers most of the region of the British Isles inhabited by the builders of Stonehenge I.

Put this way, there is no longer any strong case for the *original* decision to build Stonehenge (that is, Stonehenge I) at a special, preordained latitude. That is, I am suggesting that Stonehenge I was built as some sort of astronomical observatory that did not include lunar observations of the type made possible by the Station positions, and it was only later that the people—of a quite different culture—realized that it was a peculiarity of the location that the sight lines form a rectangle arranged with the long sides perpendicular (or as near as they could get them) to the midsummer-Sun–aligned axis of Stonehenge I, with corners set among the circle of Aubrey Holes, and pointed toward certain celestial phenomena, such as the maximum and minimum azimuths of the Moon as it rises and sets. Coupled with this is the fluke, as Hoyle pointed out, of the eclipse cycle. There are many neolithic sites broadly like Stonehenge I scattered throughout the British Isles, continental Europe, and elsewhere in the world. A later benighted culture, knowing only from their legends that this site was some sort of device for observing the heavens, may have investigated how it could perhaps behave in such a way and stumbled onto the special properties it possessed by dint of its latitude. As Hoyle suggests, Stonehenge II and III are characterized by ritual/numinous utility rather than novel astronomical observing; what could have better convinced the people that this was a sacrosanct site than the special properties that were revealed as they watched the Sun and the Moon from there, discovering the eclipse cycle fluke and convincing themselves to invest the huge amount of effort involved in erecting the sarcen and bluestone structures?

Another astronomer with whom I broached the subject of trying to interpret Stonehenge took the view that such a pursuit was pointless (which rather spoils the fun) and gave the excellent example of the Eiffel Tower. What chance would someone have in a thousand years' time, he asked, of discovering what that structure was built for if our civilization had died out and no written records were saved? I can take that argument a little further. What the future investigator might do would be to look around the world and find many basically similar structures that serve as radio masts. Finding an array of radio antennas on top of the Eiffel Tower, he would then deduce that it was built for the same purpose. That interpretation would be confirmed because, looking back from the thirtieth century, he would perceive that the tower was built at essentially the same time as radio communications were developed. In fact, we, with our more detailed knowledge of the history of the late nineteenth century, know that the Eiffel Tower was constructed in 1889 for the great exhibition in Paris, well before Marconi conducted his first experiments with radio waves. Our thirtieth-century archaeologist's deduction would therefore be bunk. My point here is that the timing is critical, and later usage does not inform one of the original motivation for an object's construction.

Given that we have dismissed lunar motion sightings, we must search for some other motivation for the design and construction of Stonehenge I, which is presumed to be astronomical in origin. There is nothing to say that this motivation was linked to the later elaborations of Stonehenge II and III, so we are not constrained by the interpretations and discoveries of the later culture that made Stonehenge what it became in the second millennium B.C. Thus we should not be constrained by the astronomical thinking that says that the conditions in the heavens were the same 5,000 years ago as they are now. All astronomical interpretations of Stonehenge made to date have tacitly assumed this to be the case. Although we know that 5,000 years ago the Sun, the

Moon, and the planets were behaving as they do now (and see later comments on the unphysical ideas of Immanuel Velikovsky), we cannot be sure that there were not some additional celestial features that are no longer seen.

Quite apart from Stonehenge, many other megalithic sites seem to have been constructed, starting around 3000 B.C., by cultures spread across the globe, having no communication with each other, but watching a common sky. Over the past few decades, the astronomical interpretation of ancient monuments—what is known as *archaeoastronomy*—has been a growing field. All manner of stone features have been investigated, from carvings and hieroglyphics to assemblages of large and small rocks, often involving buildings that admit light only at certain times of year. For example, a Neolithic passage grave at Newgrange in Ireland has a gap in its roof through which the Sun illuminates its main chamber at sunrise on Midwinter Day, or at least it did do so 5,000 years ago. That age of the grave is verified by carbon dating of charcoal fragments found therein. Why were the ancients suddenly so interested in the sky?

Obviously, the special events happening in the sky must have been short-lived phenomena (because the megalith-building phase seems to have sprung up and then receded), and we have already seen that the precession of meteoroid streams leads to periods of activity only a few centuries long. This gives us a clue. Is there any evidence, though, for the occurrence of exceptional events? If there are many streams, all with similar fluxes, then overall we might expect the year-averaged or month-averaged rates to be invariant. That would not be the case, though, if one stream were dominating the flux because a giant comet had disembogued a massive stream. The Japanese astronomer Ichiro Hasegawa has spent many years digging through ancient Japanese and Chinese records of meteors, and he has found that there were pronounced peaks in the eleventh and fifteenth centuries A.D., when the number of meteors observed rose markedly. Looking at the months of the year when these meteors were seen, he found that these agreed with the midsummer and October-November peaks to be expected for the Taurid showers. As a check to ensure that the heightened meteor count rates were not due merely to sensitization of the observers (for example, a fashion for sky watching starting up and then blowing over), Hasegawa also looked at the records of cometary observations. The naked-eye

Plate 1. The North Pole of the Moon as seen from the *Galileo* spacecraft in 1992. On a swing through the Earth-Moon system, gaining speed for its trip out to Jupiter, the *Galileo* spacecraft obtained a series of unique panoramic shots not available previously, because all previous satellite imagery was obtained from low lunar orbits. This image vividly shows the intense impact-cratering history of the Moon. (Courtesy of the National Aeronautics and Space Administration/Jet Propulsion Laboratory)

Plate 2. Barringer Meteorite Crater (commonly known as Meteor Crater), created 50,000 years ago in northern Arizona when a small nickel-iron asteroid hit the Earth. Small shards of the impactor are found in and around the crater itself, although a larger meteorite was discovered in Canyon Diablo, the sinuous riverbed snaking across this photograph. The crater is about 1.2 kilometers across. (Courtesy of Richard Grieve and Janice Smith, Geological Survey of Canada)

Plate 3. Gosses Bluff in central Australia, formed by an impact about 143 million years ago. The original crater was more than 22 kilometers wide, only this 6 kilometer-wide inner portion having survived erosion through the eons since. (Courtesy of Richard Grieve and Janice Smith, Geological Survey of Canada)

Plate 4. The Steinheim basin in southern Germany, which is 3.8 kilometers wide, was formed 15 million years ago. In the center of the crater is an uplift that is characteristic of massive impacts, a rebound frozen in the rock. To the left of the uplift lies the town of Steinheim. (Courtesy of Richard Grieve and Janice Smith, Geological Survey of Canada)

Plate 5. The pair of impact scars in Quebec, Canada, known as the Clearwater Lakes. They are 32 and 22 kilometers in diameter and were excavated 290 million years ago when a pair of projectiles—perhaps a dumbell-shaped asteroid that separated on nearing the Earth—arrived together. (Courtesy of Richard Grieve and Janice Smith, Geological Survey of Canada)

Plate 6. The 100-kilometer-wide Manicouagan Crater in Quebec, Canada, evidenced today by a circular lake. This structure was formed 214 million years ago in the Late Triassic; the environmental effects of the impact are believed to have been responsible for the Upper Norian mass faunal extinction that occurred at that time (see Figure 12). (Courtesy of Richard Grieve and Janice Smith, Geological Survey of Canada)

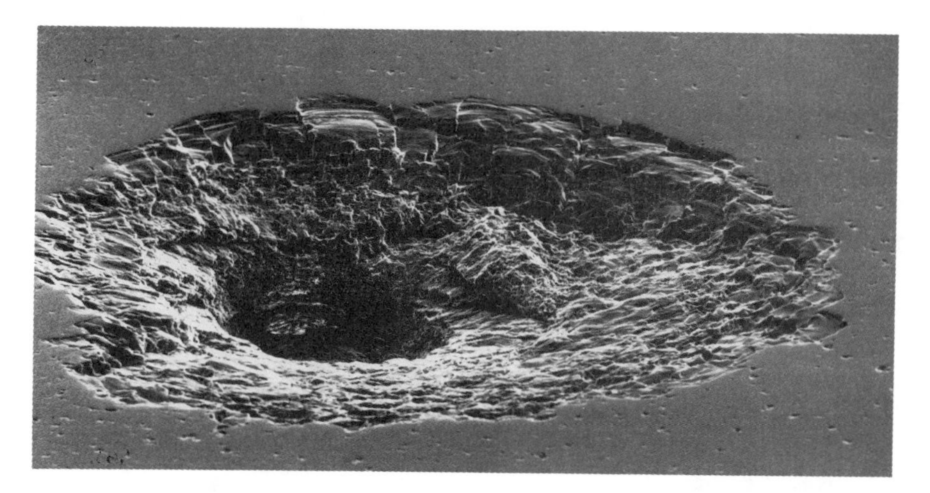

Plate 7. An altogether smaller crater: This pit, less than a millimeter across, was gouged out of one of the windows of the space shuttle *Challenger* on a flight in 1984, two years before the fateful launch of 1986. It is believed that this minute crater—which necessitated replacement of the window before the next flight—was caused by a 0.2-millimeter fleck of paint left in orbit by some previous satellite. Returned spacecraft surfaces are always pitted and punctured due to hypervelocity impacts by meteoroids, interplanetary dust, and anthropomorphic space debris. (Courtesy of the NASA–Johnson Space Flight Center)

Plate 8. Uprooted trees 2 kilometers south of the epicenter of the 1908 Tunguska explosion. Many of the trees have since rotted away, but those that remain are aligned radially away from the epicenter, as was noted by the first expeditioners in 1927. (Photo taken in 1992, courtesy of Gennedij Andreev [Tomsk University])

Plate 9. One of the tree trunks that was charred by the Tunguska explosion. This trunk, standing 1.5 kilometers west of the epicenter, was snapped off by the blast wave a few meters above ground level. (Photo taken in 1991, courtesy of Gennedij Andreev [Tomsk University])

Plate 10. The marker at the epicenter of the Tunguska explosion, as identified by Leonid Kulik in 1927. The scientist taking notes in this 1990 photograph is Yuri Kandyba, who took part in the first modern Tunguska expedition in 1957 and has been a member of many scientific parties since.(Courtesy of Gennedij Andreev [Tomsk University])

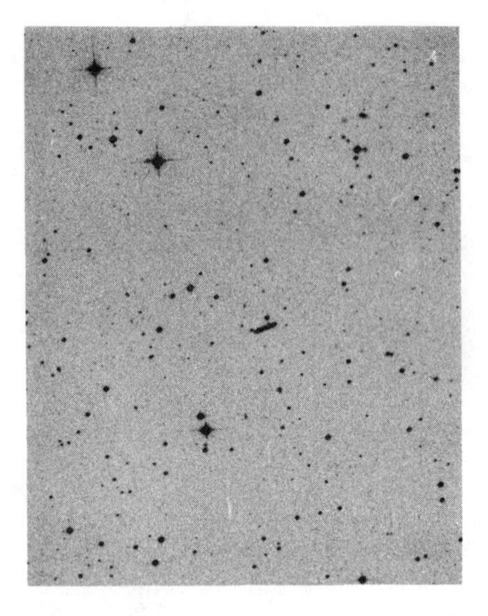

Plate 11. Against a background of more distant stars, this dense trail was captured on a three-hour exposure taken with the U.K. Schmidt Telescope in Australia. During that time, the responsible asteroid—a main-belt body called 695 Bella—moved 200,000 kilometers to produce this trail, just 2 millimeters long on the original photograph. (Copyright Anglo-Australian Telescope Board [1991], by permission)

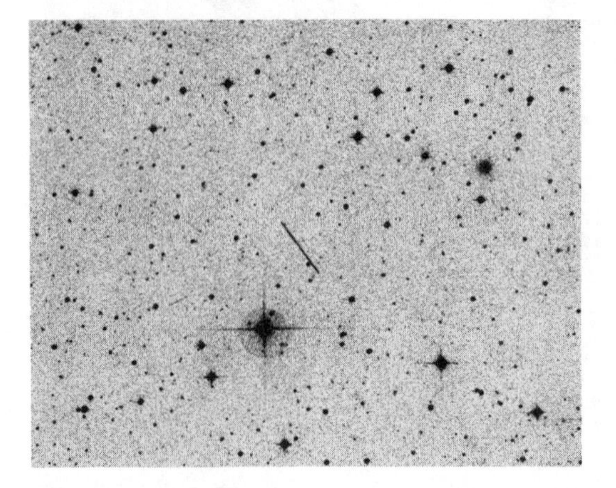

Plate 12. Two asteroid trails near one another. The obvious dark streak is the discovery detection of Mars-crossing asteroid 1991 EB_1. To its left and slightly below is the shorter, much dimmer trail due to some small, anonymous main-belt asteroid. (Copyright Anglo-Australian Telescope Board [1991], by permission)

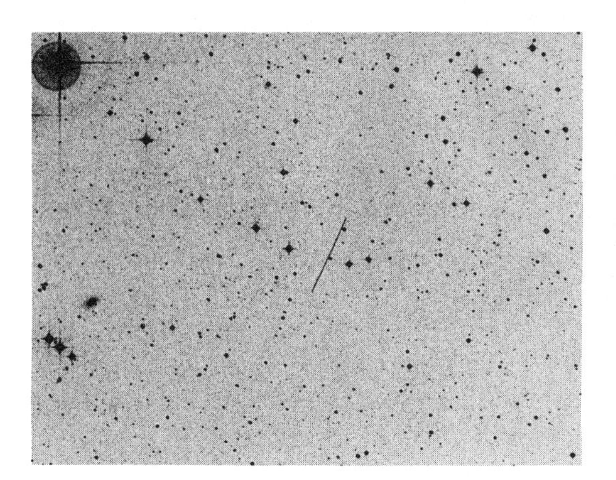

Plate 13. The elongated trail of an Earth-crossing asteroid. Such objects passing close by our planet produce longer trails because their angular speeds across the sky are higher. Apart from the various bright stars in this frame, there is also a dwarf elliptical galaxy—the diffuse image at the left. (Copyright Anglo-Australian Telescope Board [1991], by permission)

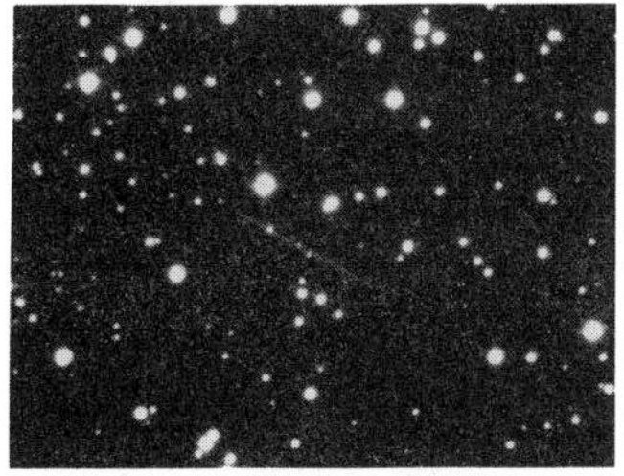

Plate 14. Astronomers are accustomed to working with negatives (and they also draw lunar maps with south at the top because that's the way it appears through a telescope). This picture rectifies matters, just to show what an asteroid trail really looks like. This is the discovery shot of 5335 Damocles, an asteroid with an orbit radically dissimilar to any other. Damocles may well be an extinct comet. (Copyright Anglo-Australian Telescope Board [1991], by permission)

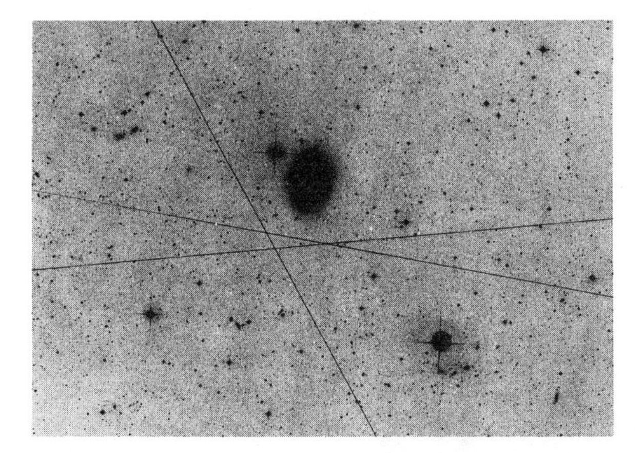

Plate 15. Not only asteroids and comets produce trails on long-exposure photographs. From this greatly enlarged image, it might be imagined that three artificially produced satellites almost collided near the galaxy NGC 4450 (the large dark, diffuse patch), although in fact they would have crossed that point at different times and at different heights. At the lower right, near the bright star, there are several tiny images of other, more distant galaxies. (Copyright Anglo-Australian Telescope Board [1992], by permission)

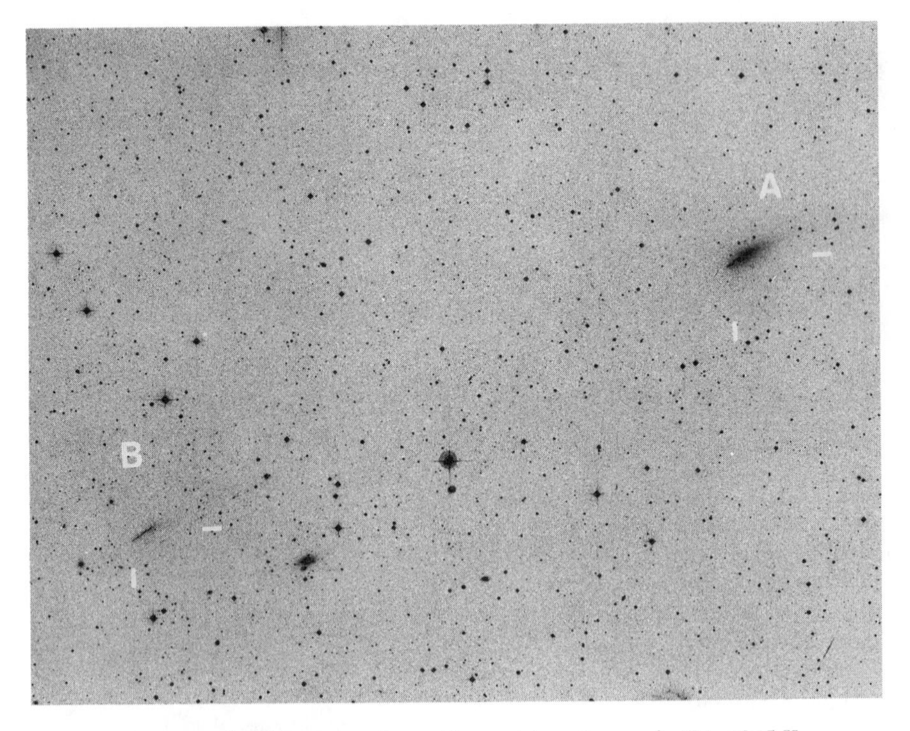

Plate 16. An example of the comet/asteroid transition. Comet du Toit 1945 II was found half a century ago but then not observed again until 1982 when Malcolm Hartley rediscovered it as a split object. The two fragments (A and B) are shown in this single negative print. In the intervening six trips through the inner solar system—it has a period of 5.2 years—it was not observed, and presumably had an asteroidal appearance, making it very faint. With the splitting, volatile materials were exposed and a bright, cometary appearance resumed. The brightest of the two fragments in this negative image returned in 1987, but the dimmer one has not been seen again. The comet is now called P/du Toit-Hartley. (Copyright Royal Observatory, Edinburgh [1982], by permission)

Plate 17. Comet Halley as photographed on March 13, 1986—the day that the *Giotto* spacecraft swept by the comet—using the U.K. Schmidt Telescope in Australia. The trail stretches more than three degrees from the nucleus in this view (six times the lunar angular diameter), although its full length could be traced for 18 degrees. (Copyright Royal Observatory, Edinburgh [1986], by permission)

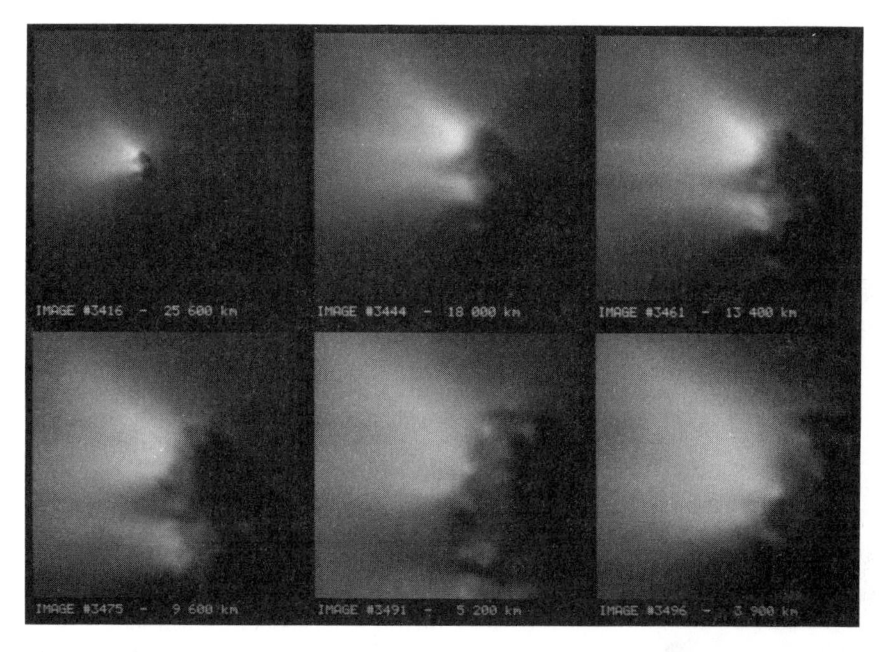

Plate 18. The only comet whose nucleus we have seen in detail. These images of Comet Halley were obtained by the camera on board the European Space Agency's *Giotto* spaceprobe as it swept past the comet on its last apparition. The nucleus measures about 15 by 10 by 8 kilometers. (Copyright Max-Planck-Institut für Aeronomie [1986], courtesy of Dr H.U. Keller)

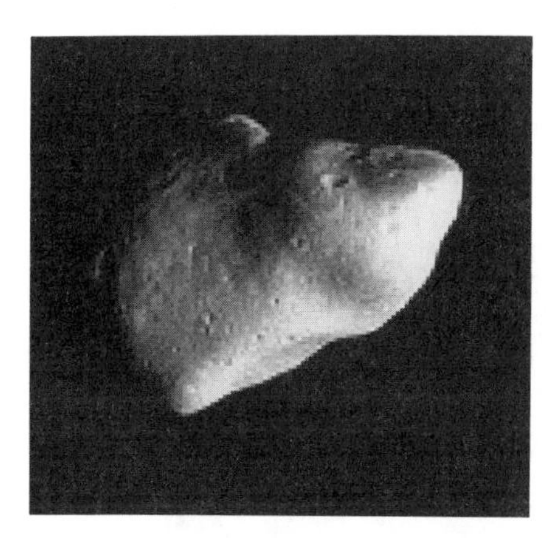

Plate 19. The main-belt asteroid 951 Gaspra as observed by the *Galileo* spaceprobe on October 29, 1991, from a range of 16,200 kilometers. The region in sunlight here measures about 16 by 12 kilometers. The smallest craters, produced by meteoroid impacts, are about 300 meters in diameter. (Courtesy of the National Aeronautics and Space Administration/ Jet Propulsion Laboratory)

Plate 20. Main-belt asteroid 241 Ida as seen from *Galileo* on August 28, 1993. This is a mosaic of five images taken at ranges between 3,000 and 3,800 kilometers, while the asteroid was about 3 AU from the Sun. Ida is about 52 kilometers long. (Courtesy of the National Aeronautics and Space Administration/Jet Propulsion Laboratory)

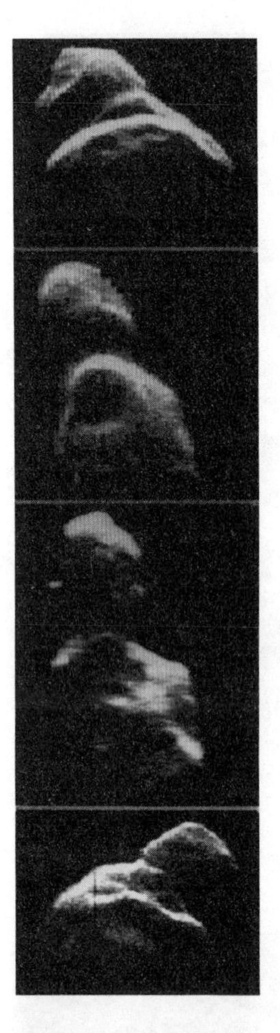

Plate 21. Preliminary radar images of Earth-crossing asteroid 4179 Toutatis obtained on (from top to bottom) December 8, 9, 10, and 13, 1992, as it passed within 4 million kilometers of the Earth. The images show two irregular bodies apparently conjoined, these being about 4.0 and 2.5 kilometers in size. The second image shows a large crater about 700 meters across. These pictures were obtained using the Goldstone Deep Space Communications Complex in the Mojave Desert of California. (Courtesy of Steve Ostro, National Aeronautics and Space Administration/Jet Propulsion Laboratory)

Plate 22. Heavily processed radar images of Earth-crossing asteroid 4769 Castalia, which show a bifurcated profile, the two lobes each being about 1 kilometer in size. For details of how these images were built up, see R.S. Hudson and S.J. Ostro, "Shape of asteroid 4769 Castalia (1989 PB) from inversion of radar images," *Science*, volume 263, pp. 940–943 (1994). Binary objects like Castalia (and Toutatis shown in Plate 21) might be responsible for twin craters such as the Clearwater Lakes shown in Plate 5. (Courtesy of Steve Ostro, National Aeronautics and Space Administration/Jet Propulsion Laboratory; copyright American Association for the Advancement of Science [1994], by permission)

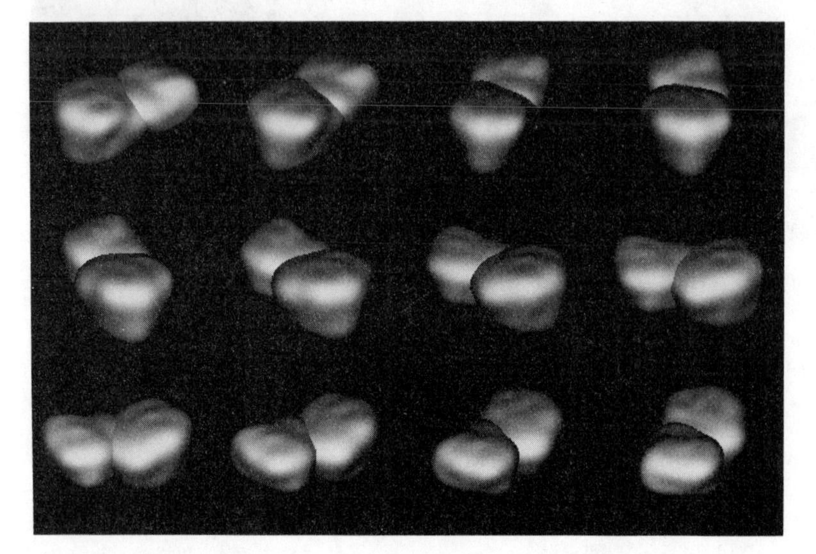

comets picked up would mostly be parabolic, for which we expect a constant flux. Hasegawa found that although there was a gradual increase in the number of comets recorded from A.D. 1 through to 1800, as might be anticipated for various nonastronomical reasons, there were no large jumps during the eleventh and fifteenth centuries. That is, it seems that the Taurids were indeed exceptionally active in those periods, reflecting precession of the stream in order to bring the core around to Earth-intercept. But what about earlier epochs?

The outrageous suggestion that I am going to make is that the Taurid Complex was producing phenomenal meteor storms between 4,500 and 5,000 years ago, accompanied by multiple Tunguska-class atmospheric detonations, and that Stonehenge I was designed to allow the (awestruck, terrified) culture of southern England to make observations of the phenomena and to perhaps predict their recurrence. Peter Lancaster Brown, in his book on megalithic sites, wrote that "Eclipses, comets and meteorites were astronomical phenomena widely observed by the ancients. But probably only eclipses were predictable. . . ." In other words, he assumed, along with many others, that Stonehenge could only have been built for eclipse prediction, comet returns not being foreseeable until Edmond Halley came along, meteorite falls also being inherently unpredictable. I am going to differ from that opinion here, and in passing note that when scientists write "probably" or "it is likely," in fact, they mean "there is a vague possibility"—a fault of which I am undoubtedly guilty in this book.

We have seen in Chapter 7 that Comet Encke is associated with the Taurid Complex. If we follow the orbit of P/Encke backward, we find that it intersected that of the Earth around 1,800 and 5,000 years ago (at its ascending node) and 2,100 and 4,700 years ago (at its descending node). This is useful because it tells us that objects with orbits like that take around 7,000 years to complete a full precession cycle, and during that time there are four epochs in which close approaches to the Earth occur, that is, parent objects can produce four meteor showers. However, we should remember that P/Encke suddenly brightened a couple of centuries ago, not having been seen previously (even though it would have been expected to have been a naked-eye comet, extrapolating its current dimming backward and returning every three years), so it seems that at least for some centuries it was of asteroidal brightness. With this in mind, we should not necessarily think of it as being the sole parent of the complex. Other large objects have also contributed, and the times at which the overall products intersect the Earth will be different. Also, P/Encke is not positioned centrally in the trail of debris detected by the Infra-Red Astronomy Satellite in 1983.

It would be useful if we could carry out precise dynamical evolution studies of the Taurid meteoroid stream by using individual meteor orbits, but because meteors are only observed in the atmosphere for a second or so, the orbits that we determine for them are always imprecise. This is especially true for the radar-detected daytime showers (about which more is stated later), but, in any case, the meteors that we have been able to observe in the present epoch are on the periphery of the stream and so do not represent the orbit of the core. Back in the early 1950s, however, Fred Whipple, in collaboration with Salah El-Din Hamid of the Helwan Observatory in Egypt, carried out a detailed study of the evolution of the nighttime meteors for which Harvard University astronomers had determined good orbits during the previous two decades. They found that there was evidence of an exceptional event, which they saw as a break-up of a proto-Encke's comet around 4,700 years ago. Due to the necessary approximations applied in Whipple and Hamid's modeling, that age should be considered to be good to within perhaps plus or minus 500 years. They also found evidence of another break-up about 1,400 years ago, except that they found the orbit of the parent object to be different from that of P/Encke. They suggested that this was a fragment of the original; that is, they modeled the origin of the Taurids as a hierarchical disintegration of a large comet, proceeding over many thousands of years. Similarly, work by myself, with David Asher and Victor Clube, and by Poulat Babadzhanov and Yuri Obrubov,[5] has shown that about 5,000 years is required for the characteristics of the four main meteor showers observed now to have been attained. Note that this does *not* correspond to the times when P/Encke had a node near 1 AU, but that is not the problem; the point is that it does indicate that some sort of cometary break-up occurred 5,000 years ago, which is precisely when Stonehenge I was being built, and the complex of material released into the stream would soon have precessed so as to intersect the Earth.

In the intervening five millennia, we might expect that three distinct periods of exceptional meteor activity would have occurred: one about 5,000 years ago (because by chance the complex formed when it had a node near 1 AU), once in the first half of the present millennium (and there are good reasons to expect it to be split into two periods, as in the eleventh and fifteenth centuries), and once between. These episodes would not necessarily be equally spaced. Clube and Napier have discussed extensively the evidence for various medieval records describing catastrophes associated with the great meteor showers of the eleventh and fifteenth centuries, and there is similar mythical and legendary evidence for Armageddon-type events over the past few millennia, in particular in the last few centuries B.C. and the fifth and sixth centuries A.D. Note that the analysis of Whipple and Hamid indicates a break-up that corresponds with the latter (that is, around 1,400 years ago). Clube and Napier have argued that the Dark Ages commencing around then were *caused* by the conflagrations resulting from intersections between the core and the Earth.

As discussed, there are four main meteor showers associated with the Taurid Complex, two nighttime showers presently occurring in October-November and two daytime showers in May-July. We have also seen that it could take up to 7,000 years for the orbit to completely precess around and produce the four distinct showers; in fact, due to their initial orientations, 5,000 years has been just sufficient. One possibility therefore would be that the parent comet arrived in its orbit about 5,000 years ago and immediately broke up, producing the complex except with later disintegrations (like that at 1,400 years ago) later adding to it. However, it could also be that the parent comet arrived rather farther back in time, and the fragmentation 5,000 years ago was long after the arrival date. One way in which that possibility could be investigated would be to investigate whether there are other showers from the same complex. The point is that the precession rate of P/Encke and the stream branches producing the four main showers implies a loop in 7,000 years. Meteoroids released earlier might have slightly smaller or larger orbits, however, and precess at a different rate. In fact, studies of meteor showers have shown that there are two other sets of four showers, usually termed the *Chi Orionid* and the *Piscid* streams, that are indeed related to the more powerful Taurids. Studies of the differential precession rates show that these required about 20,000 years to have attained their present situations; that is, the cometary progenitor arrived in the inner solar system about 20,000 years ago. The fact that these other showers are much weaker than the Taurids also implies that there must have been a later set of disintegration events, as deduced by Whipple and Hamid: If the comet had simply decayed gradually, as astronomers tend to model cometary behavior (obviously being in error to do so), the Chi Orionids and the Piscids would be stronger showers than the Taurids.

If a comet *had* arrived in the inner solar system 20,000 years ago, episodically liberating large amounts of dust, then surely we would have other evidence to back up this contention? Two sources of evidence spring to mind. First of all, the meteoroids and dust striking the Earth could affect the climate through dust-veiling of the atmosphere and would eventually settle out and leave dust deposits in the polar ice caps. In 1983, Paul LaViolette submitted his Ph.D. thesis to Portland State University, presenting the results of his study of late Pleistocene polar ice, the period from 20,000 to 14,000 years ago. This is the period when the last Ice Age was terminating, with the occurrence of abrupt climatic alterations. He found that there is a much higher concentration of iridium and nickel, apparently derived from interplanetary dust, in the ice cores at those levels than would be produced by the present influx of dust to the Earth. Further, he found that there were five distinct episodes of dust deposition during those 6,000 years, which correlates rather nicely with the expected precession rate: Four in 7,000 years is the present rate, but earlier the orbit would likely have been larger, giving a shorter precession period. During those five epochs of heightened dust influx to the Earth, the collection rate of such particles was about a hundred times as high as they are presently. This is obviously a major series of events, which would be expected to affect the terrestrial environment in one way or another.

Such dust is gradually slowed down in the Earth's atmosphere unless the particles are large enough to ablate, in which case their evaporated products will eventually settle out. The Moon possesses no such atmospheric shield, however, so dust and meteoroids alike strike the lunar surface without being decelerated. The second source of evidence for the interplanetary dust flux being higher in the past than it is now comes from the tiny craters and pits produced by their impacts. Studies of lunar rocks returned during the Apollo program allowed the ages of such microcraters to be determined in the following way. Rocks exposed on the lunar surface are continually bombarded by charged particles from the Sun (solar cosmic rays). These penetrate the rocks and leave tracks in the crystalline structure which can be counted, with the density of tracks being an indicator of the exposure age of the rock. The same method is used for determining the length of time between a meteorite's release from its parent and its arrival at the Earth. In the case of a microcrater, however, the clock is reset: As the dust grain impacts and excavates the tiny pit, it melts the rock in the base of the crater, obliterating any earlier tracks. Thus, by counting the cosmic ray tracks in that melted rock, the time since the formation of the microcrater can be determined.

One of the major players in this work was Herb Zook of NASA–Johnson Space Center in Houston. In collaboration with others, he set about determining the age distribution of these microcraters, but he came up with a surprising result: He found that the impact rate from 20,000 years ago was about ten times as high as that expected from the present population of dust. Zook is a fine scientist, so he took a look at the assumptions upon which his analysis was based, and in particular he noted that he had assumed that the solar cosmic ray flux was the same 20,000 years ago as it is now, implying that the rate of track production has been invariant over that time. There were two possible interpretations of the data, which he discussed at length:

1. The Sun was much more active 20,000 years ago, with more solar flares occurring, these producing a higher flux of solar cosmic rays;
2. There have been one or more great episodic enhancements in the interplanetary dust population over the past 20,000 years, likely resulting from one or two very large comets arriving in the inner solar system.

Zook decided that item 1 was the more likely answer. I believe that Zook made the wrong choice. For one thing, there is no other evidence to support a conjecture that the Sun was much more active so (astronomically) recently. I interpret the lunar microcrater ages as being evidence in support of the idea that a giant comet broke up in the inner solar system, starting around 20,000 years ago. One can also appeal to other direct measurements of the exposure ages of interplanetary dust grains collected in the upper atmosphere: The same cosmic ray track method renders an exposure age for these of 10,000 to 30,000 years.

There is contemporary evidence for the zodiacal dust clouds having been boosted by the Taurid Complex. Earlier I wrote about the dust detector on the *Giotto* spacecraft operated by Tony McDonnell and his team at the University of Kent. This was an active detector, which counted the impacts on the spacecraft as it flew through the dust could surrounding Comet Halley. However, McDonnell also had a passive experiment on board NASA's Long Duration Exposure Facility satellite (LDEF), which was in Earth-orbit from 1984 to 1990. That experiment—actually a series of detectors, on different faces of the LDEF—consisted of thin sheets of metal foil, which would be cratered by very small dust particles and punctured by larger ones. When LDEF was returned to Earth and dismantled, researchers at the University of Kent began counting the crater pits and perforations, using high-powered microscopes and other techniques. The result was a surprise: They had expected the numbers of impacts on opposite faces—in particular, the face always pointing north of the ecliptic and the one directed southward—to have equal numbers of strikes, because one might expect there to be as much dust passing its ascending node as its descending node. This would have provided a useful bench mark for the interpretation of the data from the other faces. The north and south faces did not play ball, however. There were many more dust impacts on the north face than on the south. Neil McBride and Andrew Taylor, two young researchers at Kent, started to scratch their heads and look at plausible reasons for this being the case. They knew that another dust impact detector on LDEF, an active counter that had worked only for the first 9 months of its lifetime in orbit,[6] had produced evidence of enhanced dust impact rates during meteor showers—and that was not expected, because the conventional wisdom says that small dust grains are absent from meteoroid streams—so McBride and Taylor looked at whether discrete meteor shower effects could produce the north-south asymmetry. They were having great difficulties with this when I pointed out to them that the subtle effect they were looking for might be accommodated by the Taurid Complex, in particular because it is symmetric about Jupiter's orbital plane rather than the ecliptic. Of course, I had in mind the interpretation given here of the lunar microcraters, and also Fred Whipple's suggestion back in 1967 that the Taurids and Comet Encke are powering the zodiacal dust cloud. As the reader has likely anticipated, McBride and Taylor found that the LDEF data are fitted rather nicely by a model using a main source based on the Taurids.

Let us return to the larger products of such a comet disintegration. From a catastrophist viewpoint, the association of huge atmospheric detonations similar to Tunguska with intense meteor showers does not appear ludicrous at all, and yet such ideas have been dismissed with vigor from mainstream science. The culprit—or maybe we should simply say the catalyst—was Immanuel Velikovsky. In a series of books published around the middle of this century, the most notorious of which is the misnamed *Worlds in Collision*, Velikovsky put credence in the various historical records and mythical depictions (including many in the Bible) of a huge comet crossing the sky with attendant calamities and came up with the absurd idea that the "comet" was in fact Venus having a near-miss of the Earth. Anyone with knowledge of college-level physics should be able to work out for themselves that Velikovsky's idea is in breach of various laws of physics and hence is untenable. Nevertheless, a breed of Velikovsky disciples emerged, similar to alien-contact enthusiasts, and they proved to be the bane of astronomers, with occasional resurrections occurring even today.

The real problem for science is that astronomers, in America in particular, became so entrenched and vehement in their criticism of Velikovsky's astronomical nonsense that their mindsets also became instilled with not only a rejection of, but also a nonconsideration of, the possibility that the myths and records of past civilizations might contain important information about what was happening in the sky in pre-modern times. In fact, the similarity between the legends of disparate human cultures are startlingly similar. In scientific publications I have pointed out that Australian Aborigines and New Zealand Maoris have oral traditions of strange rocks falling from the sky, causing awful fires and many deaths, and this scenario is common in the myths of other peoples. On one hand, astronomers have prided themselves in instructing geologists that impact catastrophes were responsible in part for the shaping of the planet, but on the other hand, they have been blind to the fact that they have made a uniformitarian assumption when it comes to their own science: That the sky as it is now, is as it ever was, at least while humans have walked the Earth. There is ample evidence not only from historical records of various forms, but also from the analysis of data from this century (such as Whipple's modeling of the Taurid meteors), that around 5,000 years ago the sky did not appear as quiescent as it does now, and that since that time there have been other disruptions of the heavens, producing conflagrations here below.

But this chapter is not about Velikovsky; it is predominantly concerned with Stonehenge. We have seen that there is prima facie evidence that the Taurid showers were producing spectacular celestial displays, along with atmospheric detonations, at the time that Stonehenge I was built. I am suggesting that the two were linked. How does the design of Stonehenge I correlate with the showers?

When I first considered this idea, it was prompted by the recognition that the two daytime meteor showers of May–July, which derive from the Taurid stream, have radiants very close to, but a little to the west of, the Sun.[7] Because the stream is of very low inclination, the radiants are close to the ecliptic (that is, close to the track taken by the Sun across the sky). Thus in the few weeks straddling midsummer, if the stream were producing displays far more intense than those occurring now, one might see many meteors emanating from the northeast, starting about an hour before the Sun rises. As the sky brightened, it would drown out the display of shooting stars, although in fact their flux into the atmosphere above your head would continue to grow for another six hours or so. The fireworks would therefore herald the rising of the Sun, and during that day the concussions from multiple Tunguskas would terrify the people below—sounds like a lot of myths and legends to me, and indeed religious beliefs connected with the coming of Armageddon!

Nice idea though that may sound, is there any truth to it? I knew that Gerald Hawkins, who had done so much to promote the lunar/solar eclipse concept for Stonehenge in the 1960s, had started his career in radar meteor research at Jodrell Bank in England, where the May-July daytime showers were first studied, and even while he was writing his book on Stonehenge, his main research was on meteors. Could he have missed something so obvious? The answer was that I had been much too simplistic: Although the daytime Taurid showers are active (to some extent) in the present epoch, they would not have been peaking in the summer when Stonehenge I was built. The dates and the radiants of the Taurid Complex showers were not the same 4,500 to 5,000 years ago, due to precession—the precession of the equinoxes and the precession of the stream orbit. By taking a slight diversion, we will find out why.

It is a fallacy that there will not be a February 29 in the year 2000. How many people believe there will not be, I could not guess. The source of the fallacy is the mistaken belief that all centurial years are *not* leap years, despite the fact that they are divisible by 4. This was the case for the year 1900, but it will not be true in 2000 because there is another part to the rule: A year *is* a leap year if it is divisible by 400.

But why does such a rule exist? As you might imagine, it does have something to do with astronomy. If it took the Earth exactly 365 days to complete each orbit, we would not need leap years at all. If it took exactly 365.25 days to complete an orbit, we could have a leap year every fourth orbit, as regular as clockwork. But the actual length of the year, astronomically speaking, does not "play ball." The first problem is defining what we mean by a *year*. There are a number of definitions, each having a different duration.

The time for the Earth to pass from perihelion on one orbit to perihelion on the next is called the *anomalistic year*, and to the nearest tenth of a second it is 365.259635 days long. Precession of the Earth's orbit under the influence of the other planets, however, means that this is not a useful definition of the year for civil use. *Sidereal time*, which merely means the time according to the stars rather than according to the position of the Sun in the sky, is the time system used most often in astronomy. A sidereal day is actually shorter than a normal day by about one part in 366 (and a bit) because the Earth spins on its axis 366 (and a bit) times in every year—once for each day, but also once due to its orbit around the Sun. The time that it takes the stars to come back to their original positions is called a *sidereal year*, and it is 365.256363 days long. Again, this is not a useful measure for civil use.

What *is* useful to people in general is a year counted between the equinoxes. What is an equinox? As the Latin derivation of the name suggests, it is the time of year at which night and day are the same length. This occurs precisely when the Sun in the sky appears to cross the plane of the Earth's equator, this being about March 21(the vernal equinox) and about September 22 (the autumnal equinox). The time between vernal equinoxes is 365.242190 days. This is the standard by which we determine a year in the civil sense; if we did not keep a year to that average length, then over the centuries the seasons would gradually slip until the northern-hemisphere summer would start in December. This definition of a year is called the *tropical year*.

If a year is 365.242190 days long, then the frequency of leap years must be adjusted to compensate. For reasons that will become apparent, let us take 400 years as our baseline. That must last for 400 × 365.242190 = 146096.876 days. If we had a leap year every 4 years, then 400 years would occupy 146100.0 days, which would mean that we were three days and a bit ahead. If we missed out *every* centurial year, we would have 146096.0 days, being almost a day short. If we had a leap year additionally in every century year which is also divisible by 400, however, then we would have 146097.0 days in every 400 years, which is just 0.124 days out; that is, about three hours over 400 years. Thus, with the present system of leap years, after about 1,600 years we will be close to half a day out and an adjustment might be necessary, but we may leave this for our descendants to worry about.

The calendar that had every fourth year as a leap year, the Julian calendar, was introduced by Julius Caesar in 46 B.C. This worked well, except that time gradually slipped until by A.D. 1582 the calendar was out of step with the seasons by 11 or 12 days. Pope Gregory XIII therefore decreed that leap years should not occur in centurial years unless they were divisible by 400, thereby introducing the Gregorian calendar, which we still use today. Gregory also decreed that the date should be changed in order to get rid of the 11 days that had slipped. This edict resulted in riots because some people thought that 11 days were being stolen from their lives.[8]

The purpose of this discussion is not solely to enlighten you as to why A.D. 2000 will be a leap year, but mainly to point out that there is a difference of 0.014173 days *per annum*[9] between the sidereal year (according to the stars) and the tropical year (according to the seasons). Thus there is a slippage between them of close to 14 days per millennium, this being termed the *precession of the equinoxes*. This would mean that, all other things being equal (which they are not, as we'll see), a meteor shower occurring on a specific date in the late twentieth century will occur a week later in the (tropical) year in the late twenty-fifth century.

The thing that is not equal, as mentioned in the last paragraph, is that apart from the date of a meteor shower being shifted by the precession of the equinoxes, precession of the meteoroid stream orbit caused by planetary perturbations will also alter the date on which a shower occurs. For example, the Taurid stream precesses at a rate of six to eight days per millennium, depending critically on the semi-major axis and eccentricity of the particular orbits under consideration. By chance, the rate of precession of the Leonid stream is virtually identical to the precession of the equinoxes, both being 14 days per millennium, meaning that a thousand years ago the Leonid meteor storms would be expected to have been occurring near mid-October rather than mid-November, and this is borne out by historical records. Note that the precession of the streams means that the shower radiants would be in different constellations from those occupied currently.

What does all of this mean for the Taurids? When I applied the appropriate precession rates to the daytime showers, I found that back in 3000 B.C. they would have peaked in early March, 110 days earlier than they do today; that is, 22 days earlier per millennium. Clearly, this could have nothing to do with the main axis of Stonehenge, because the daytime showers would be active near the Spring Equinox when the Sun rises due east rather than in the northeast.[10] But what about the nighttime showers? Although these are currently active from mid-September through early December, the Southern Taurids peak around November 3, the weaker Northern Taurids about ten days later. These are soft peaks, however: The low inclination of the stream leads to shower activity being sustained for many weeks. Around 5,000 years ago, one would expect the Southern Taurids (as observed now) to peak 110 days earlier, in mid-July, but with the activity starting around midsummer. If one used instead the best-available determination of the orbit of the Northern Taurids (the other nighttime shower), one derives a date of peak activity in the last week of June. Thus it is entirely feasible that the core was intersecting the Earth 5,000 years ago near midsummer, producing spectacular nighttime meteor storms with accompanying conflagrations and calamities.

We have seen previously that the daytime showers have radiants very close to the Sun. It is simple to see that for the nighttime showers the radiants are in precisely the opposite direction to those of the daytime: close to the antipodal point of the Sun (the point that astronomers call *opposition*), but an hour or so to the west of it.[11] Sunrise is close to 3:50 A.M. around midsummer at Stonehenge, so the shower radiant would rise at about 5 P.M. while there is still around four hours of daylight left. Due to considerations of the projected target area of the atmosphere, the meteor rate (for both large and small objects) would gradually rise as dusk passed, continuing to grow until the radiant reached its highest point in the sky, reaching a crescendo around midnight which would continue for the next three to four hours until dying away as the Sun rises and drowns out all but the brightest fireballs visible in daylight. No wonder the ancients were so keen to see the Sun rising, sweeping away the terror of the night! This desire to see the Sun rising provides an explanation for the alignment of the main axis toward the northeast, the Heel Stone and its twin forming a gap through which the Sun would appear.[12]

The absence of any written records from the Stonehenge people(s) means that this is mere supposition (but so are all interpretations). However, written material was passed down in cultures elsewhere which would have experienced the same celestial portents and terror, and a suitable perspective allows one to make sense of what was written. For example, the Greek philosopher Plato, writing his book *Timaeus* in the fourth century B.C., related information from earlier Egyptian records which spoke of

> . . . a deviation of the bodies that revolve in heaven around the Earth and a great destruction, occurring at long intervals, of things on the Earth by a great conflagration. . . . once more, after the usual period of years, the torrents of heaven will sweep down like a pestilence, leaving only the rude and the unlettered among you.

This sounds like what might be expected should the hypothesis be correct that the Taurid stream caused phenomenal meteor storms with associated Tunguska-type explosions and blast waves. But what of "the usual period of years"? Remember that the smaller meteoroids would be distributed around the stream orbit so that a shower would be expected every year, and quite likely a spectacular one. But the real storms, along with the detonations, would happen only when the Earth passed through the concentration of large bodies in one region of the stream, in a way similar to the Leonid storms occurring on a 33-year cycle today. For the

Taurid stream, with an orbital period of about 3.3 years, the cycle of storms/detonations would be 10 years or thereabouts; the group of large objects would miss the Earth on two out of every three passes through our orbit, but hit on the third. Alternatively, it would be simple to hypothesize that the meteor storms occurred every 19 years, to fit in with the apparent cycle that Stonehenge III follows. For example, if the core of the stream and the parent had a semi-major axis of 2.13 AU, and hence an orbital period of 3.1 years, on every sixth orbit it would pass by the Earth, by chance fitting the 18.61-year lunar cycle. As Lancaster Brown writes, what is required for any interpretation is "a possible spectacular event performed by 'something' at Stonehenge every nineteen years." There are many plausible Taurid Complex orbits that could be proposed in order to fit a cycle of 19 years or thereabouts.

Before passing on to Egypt, we should finish our discussion of Stonehenge by at least making suggestions for the purpose of all of the salient features of Stonehenge I. We have already seen that the main axis is aligned with where the Sun rose on Midsummer Day, seeming to chase away the meteors whose radiant had risen nearby about 11 hours earlier. The Heel Stone and its twin, which may or may not have been present, mark that axis. Apart from the gap producing into the Avenue to the northeast, the circular bank and ditch of Stonehenge I was complete, forming an enclosure, except for a causeway at the south through which people entered. Although that bank is now barely shin-high, when it was built it was just over head-high. Its purpose could have been to form a level, artificial horizon for observers inside, but an alternative explanation might be that it was a light baffle. When people ask me how to observe shooting stars, the first thing I tell them is that they need to go somewhere with dark skies, far from city or house lights, and after they start looking, they need to allow 10 or 15 minutes for their eyes to become dark-adjusted. Any disruption—for example, brief use of a flashlight—is enough to destroy that dark-adjustment. Although the ancients did not have city lights or automobile headlights to contend with, a glimpse of a campfire or burning torch, their only sources of light at night, would "flash" one's eyes and ruin that individual's observations for a while. This might not have been a problem at times of strong shower activity, but if the point of Stonehenge I was to monitor meteor rates in order to predict when storms were due, obviously a dark, protected area would have been necessary. Currently, astronomers seek sites for observatories far from artificial light sources[13]; 5,000 years ago, all they needed to do was to build a suitable head-high bank.

The final feature of Stonehenge I is the circle of 56 Aubrey Holes. I am not sure what to make of them; but then, nor does anyone else, although there have been many suggestions. They are sometimes interpreted as providing a network of sight lines around the horizon (inside an obscuring bank?), but Hawkins and Hoyle both hypothesized that they were year or month counters of some description. For the sake of argument, I will give an alternative explanation. When I started out in meteor research, I recall reading a book on the subject which contained a photograph of Canadian meteor watchers. There were about ten of them, all laid down and arranged in a circle with heads pointed inward and feet outward. The startling thing was that they were all enclosed in what looked like individual coffins, with an open area of the lid such that they could look up at the sky. Especially in the winter, it gets damned cold at night in Canada, and these "coffins" were well-insulated cocoons of heat. Perhaps the Aubrey Holes, which were just the right size for a human to sit in, performed a similar function. It may not be *that* cold in England during summer nights, but remember that this was before the invention of polyfill sleeping bags, and the poor observers had to be in the open and distant from their normal heat source (campfires).

And now to Egypt. Let me apologize in advance to anyone expecting a discussion of the pyramids with detail similar to that given for Stonehenge. The following discussion is just to hazard a few speculations about these other famous archaeological relics, which the reader might find of interest.

Earlier we mentioned that some of the first serious work on the astronomical orientations of Stonehenge was done by Sir Norman Lockyer in the first few years of this century, although the alignment with midsummer sunrise had been noted long before. In fact, Lockyer cut his teeth in archaeoastronomy on the area of Egypt, and he had spent some time studying the orientations toward celestial events of many of the Egyptian temples. Later work has confirmed some of his ideas. In particular, it is known that many Egyptian temples admit sunlight through apertures in order to illuminate certain positions only at set times of year, in a way similar to the Newgrange chamber mentioned earlier. An example is one of the temples dedicated to Ramses II at

Abu Simbel, which dates from around 1200 B.C.; I am unsure whether the correct orientation was preserved when the temples were moved for the building of the Aswan Dam in the 1960s.

But what I want to discuss in exclusion are the pyramids, which have often been suggested to have orientations aligned with various stars[14] (for example, see the book by Lancaster Brown), and, along with the Sphinx, are literally "oriented" (that is, they face due east, the direction of sunrise at the equinoxes). The first thing to note is that the Egyptian word for *pyramid* apparently means "place of ascent", the ascending being done by the dead king entombed therein as he was transported into the sky to spend the afterlife with the Sun god, Ra. Many kings were buried with boats, because Ra was believed to cross the sky in a boat, and it was assumed that the dead king would need a similar form of locomotion. Many of the Egyptian pyramids were stepped: Indeed they were viewed as being the stairways to heaven. The development of the pyramids, of which there are 35 in Egypt, began some time after about 3000 B.C. and culminated in the colossal structures known as the Khufu (or Cheops) and Khafre pyramids, which were built at Giza, near Cairo, between 2600 and 2500 B.C. After that, the building of such gargantuan tombs went into decline, with ever-smaller pyramids being undertaken. This would fit in with a hypothesis that their construction, a phenomenal task that is still not understood from an engineering perspective, let alone a sociological one, was prompted by celestial portents lasting for only a few centuries. The later (inferior) pyramids are significant in the so-called Pyramid Texts, a set of magic spells dating from prehistoric times, giving the recipe for a safe passage by the king into the sky and his acceptance by Ra.

But why is a pyramid shaped that way? We have earlier discussed the zodiacal light, a diffuse glow in a huge triangular shape which follows the ecliptic across the sky and which is due to dust grains in space scattering sunlight. I also wrote that it is best seen from in or near the tropics, up to a few hours after sunset or before dawn; indeed the zodiacal light was known to the ancient peoples of the Middle East as the "false dawn." At such times, it may stretch far above the horizon, more than half the way to the zenith. From an especially dark viewing site, there may also be visible a dim band reaching right across the sky along the ecliptic. This *zodiacal band* is due to dust that is exterior to the terrestrial orbit, while the zodiacal light proper (the pyramids in the sky) is due predominantly to sunlight scattered from dust interior to our orbit. If one makes precise measurements, one finds that the pyramid of light and the band are not quite symmetric about the ecliptic, but are generally arrayed about the plane of Jupiter's orbit. Because Jupiter's inclination to the ecliptic is only slightly more than a degree, however, this makes little difference.

Now recall that when a large comet breaks up, a great replenishment of the zodiacal dust cloud will occur, making the zodiacal cloud and band much brighter than they appear to us today. In fact, they would look like the river that Ra navigates his boat along each day, followed also at night by the newly bright comets in low-inclination orbits, such as P/Encke. At each end of the "river" lies the triangular profile of the main zodiacal light, a pyramidal shape. It was this that the dead king needed to ascend, climbing his stairway to heaven.

Such a gross boosting of the dust cloud is indicated by both the samples from polar ice and the lunar microcraters, which were discussed previously. Early accounts of the sky from when written records are first available also speak of a phenomenon that historians, primed by astronomers who assumed in error that what was seen then was the same as what we see now, have interpreted as being the Milky Way. The same records state that this "Milky Way" was the path formerly taken by the Sun (that is, the ecliptic) and that it was produced by comets. That is obviously not the Milky Way that we see now; what was being described was a super-intense zodiacal light and band.

Another important point is how the zodiacal light and band would behave as sunrise approached. The zodiacal band would come closer to being perpendicular to the horizon as the middle of the night passes, and as the pyramid of light begins to peek above the horizon hours before daybreak, it is still tilted far over. As time progresses, it straightens more and more, never quite reaching the perpendicular position. Then the whole sky would begin to brighten and redden in the northeast, until eventually the Sun would rise from the midst of the zodiacal cloud. This would surely have created a very strong impression on the Egyptian people, especially if the zodiacal light were much brighter 5,000 years ago than it is now.

But how might the zodiacal light vary in brightness as the Earth passed through a meteoroid stream? The answer to this question was answered in the early 1970s when a French team used a satellite detector to study how the zodiacal light varies in intensity. One of the difficulties in studying the zodiacal light from the ground is that one is confused by many other light sources: stars and planets themselves, light from those objects scattered by the atmosphere, moonlight similarly scattered, the aurorae, and also light continuously emitted by atmospheric atoms (the airglow). In view of these other light sources, it is necessary to put a detector above the atmosphere if you are to make useful measurements of the constancy or otherwise of the zodiacal light. The French group found that, contrary to what they expected, the intensity jumped occasionally—doubling in one case—for a few days at a time, and those jumps recurred the next year and could be correlated with transits through known meteoroid streams. The ancient Egyptians would have seen a much more pronounced brightness enhancement due to the huge amount of dust then recently released in the Taurid Complex.

Is there anything to link the ancient Egyptians with such calamitous events occurring in the sky that have been described earlier? Well, one example is from the Egyptian hieroglyphics; the symbols for *thunderbolt* and *meteorite* are the same, and contain a star. That is, it seems that the Egyptians associated meteors and meteorites with explosions above their heads, which is certainly indicative of a tumult taking place in the sky of a type different to our experiences today. Various European cultures also associated certain gods with both thunder and meteorites—for example, Thor and Zeus. We can either shrug such things off as the imaginings of ignorant primitive peoples—who were nevertheless able to construct magnificent structures such as Stonehenge and the pyramids and orient them to various celestial phenomena—or start taking a rather different view, perhaps more enlightened, and interpret their relics in light of an alternative perspective of what might have been happening in their skies.

An appeal to disparate cultures may again be made for support for this idea. Where else in the world were pyramids built? Those that spring to mind are in Mexico (and the Mayan pyramids are known to have had astronomical/calendrical motivations), east Asia, and the Babylonian and Assyrian ziggurats. What these all have in common is their latitude—all near the Tropic of Cancer. As we know, from tropical latitudes the zodiacal light is most impressive, and therefore we might anticipate finding buildings there that mimic its shape, even though an enhanced dust cloud would still have been obvious from Stonehenge and similar northerly latitudes.

If the conjecture that the Taurid Complex produces episodic bombardment of the Earth is correct, then this has wide-ranging implications for our future. The good news is that, if our sums are correct, the next epochs of core-intersection are not due for some time, perhaps not until about A.D. 3000; nevertheless, stray objects such as the Tunguska projectile will strike home from time to time. Nothing like the phenomenal episodes that seem to have occurred after the fragmentation event around 5,000 years ago is expected, however; then again, maybe the fragmentation of a yet-undetected 5-kilometer dormant comet is imminent. Earlier in prehistory, soon after its injection into an inner solar system orbit around 20,000 years ago, the cometary progenitor certainly seems to have played havoc with the amount of dust in interplanetary space, which is not a surprise. Earlier in this book I explained that only a few cubic kilometers of cometary material, broken into grains, would boost the interplanetary dust population by a large factor.

But to close this chapter, let's note that this complex of interplanetary material is not all bad. The arrival of this enhanced dust flux at the Earth coincided with the end of the last Ice Age. This may be a coincidence, but I would be happy to argue to the contrary. Perhaps we have the Taurid Complex to thank for our present pleasant climes and indeed the chance for our civilization to ascend and flourish.

I will go further, but at the risk of inducing apoplexy in those who refuse to accept any link between astronomy and the megalithic structures we've considered. One can understand that communications between the peoples of Europe could lead to the construction of many similar stone circles or alignments of menhirs at around the same time and contemporaneous with a shift from a nomadic to an agrarian lifestyle. One can argue that the new-found production of an excess allowed our predecessors the leisure to spend erecting their megaliths. One could also argue to the contrary, however, that nomadic indolence was abandoned due to a perceived imperative to build permanent structures with which to observe and predict celestial phenomena affecting their lives on Earth. Cultures elsewhere, which certainly had no contact with the peoples of Europe, seem to have undergone a similar transition starting around 3000 B.C.—for example, the Mayans, who began farming and constructing their first permanent buildings close to that time. A common cause is required to explain these unless one falls back on the unsatisfactory possibility of pure chance. Worldwide climate change would be one possibility, but it is not evidenced in the physical record from 5,000 years ago. I suggest that something was happening in the sky, with regularity, that terrified the people in diverse and disconnected cultures, leading to a lifestyle change induced by the perceived necessity to achieve certain things: the construction of megaliths connected with what was happening far above their heads but directly affecting their lives. In more recent times, we have seen similar phenomena in human societies, with nations re-gearing their industry and their daily lives to meet the demands imposed by warfare. In times of war, huge scientific and technological leaps are achieved. A good example of parallel but independent development is the effort by both the Allies and the Nazis to build atomic weapons during World War Two; although the physics was commonly known prior to the war, it was only the stringencies of the conflict that led to the actual development of the nuclear bomb. My suggestion is that similar stringencies, due to a common warfare against the sky, led to the contemporaneous advances in many disparate cultures spread around the globe and eventually to the state of our civilization today. (The anthropologists and antiquarians may now have their paroxysms and apoplectic attacks!)

9

False Security: The Atmosphere's Shield

Despite the fact that every day around 100 tons of meteoroids cascade down on the Earth, at speeds between about 40,000 and 260,000 kilometers per hour, we humans are not continually peppered with these projectiles. Protecting us from this otherwise lethal rain is the atmosphere, which acts as a sort of shield, at least for these smaller daily arrivals. Apart from this continual stream, there is also the occasional large object, a million- or billion-ton asteroid or comet, which the atmosphere cannot protect us against. In some ways our atmosphere could be thought of as acting like a sieve rather than a shield, but a sieve that lets the larger particles through while sifting out the smaller ones.

Sieve or shield, the question is: "Just how good is the protection we get?" An umbrella will stop the rain from soaking you, but will it save you from a brick dropped off the top of a highrise building? Earlier we saw that a 1-meter-square column up through the atmosphere has a mass equivalent to a rocky boulder about 4 meters in diameter. This provides a baseline minimal guess for the size of an object that might be required to punch its way through to the surface, with 8 or 10 meters being closer to reality due to various considerations. The reasoning here was very simplistic, however, and obviously the minimum size that might reach the surface intact depends on several factors, such as its speed, angle of arrival, density, composition, and so on. If you do a belly-flop into a swimming pool from the side of the pool, because you are not moving quickly, it does not hurt a great deal; but a belly-flop off the top diving board can sting a great deal because your speed is high as you hit the water, and it is painful even though water might be thought of as being a soft substance (a fluid). Similarly, we might expect that a higher-velocity (say, 60 to 70 km/sec) object meeting the top of the atmosphere will be less likely to survive to the surface than a low-speed (say, 12 to15 km/sec) object, all other parameters being equal. Those other parameters affect the object's survival probability in similar ways.

Meteors, or shooting stars, are produced when small solid bodies enter the atmosphere at the high speeds they must possess due to their orbital motion about the Sun and are decelerated by the uppermost layers of the Earth's gaseous shroud. If the atmosphere were many thousands of kilometers thick, the meteoroid would be gradually slowed down and would remain intact, eventually falling to the ground at the same speed it would if it were dropped from an airplane—maybe 200 kilometers per hour (kph), depending on its density and cross-sectional area. In reality, however, the terrestrial atmosphere is quite thin (using *thin* in the sense of linear thickness, as opposed to using the word *tenuous*). At an altitude of 100 kilometers, the air density is about one millionth that at the surface, and that density drops off by about a factor of 2 with every extra 5 kilometers in height. An incoming particle will start to feel an appreciable drag force at about 150 kilometers' altitude, as it strikes more and more air molecules. A meteoroid, say 1 centimeter across, will have been stopped by that drag by the time it gets to half that height (at 75 km). By then, it has met more than its own mass of atmosphere. If the meteoroid was originally traveling at 30 km/sec and arrived at an angle of about 45° to the vertical, it would take three to four seconds to pass from 150 kilometers' altitude down to 75 kilometers.

In that amount of time, it would need to lose all of its kinetic energy (its energy of motion), which we know is equivalent to about a hundred times the chemical energy of its own mass of TNT. As the meteoroid is slowed down, it begins to heat up, and this will lead to its radiating away some of that heat as infrared energy. It could also lose a little heat from conduction to the air molecules that it meets, heating them up. In three or four seconds, however, there is just not enough time for the large amount of energy that it possesses to be radiated or conducted away. Instead, the heating of the meteoroid continues unabated, and within the first second it has begun to melt. There is not even time for the heat to be conducted into the center of the meteoroid, and the outer layers start to boil off, carrying heat away with them as they evaporate. This process is called *ablation*. As the meteoroid burrows deeper and deeper into the atmosphere, more and more of it ablates away, until by 75 kilometers there is practically nothing left.

You witness the meteoroid's death throes as a shooting star because the atoms of the solid body glow as they boil off, also exciting the neighboring atmospheric gases to glow briefly. The gases emitted by an ablating meteoroid are at temperatures of 3,000°C to 10,000°C, so it is not surprising that they can be seen so far away. In fact, so much energy is released during the ablation that the outer electrons are stripped off of the atoms, leaving a linear trail of charged atoms (ions) and electrons, which can be detected with a suitable radar, as mentioned earlier. It is using such radars that the influx of particles with masses below about one hundredth of a gram has been determined and daytime meteor showers studied: Meteor radars[1] are not affected by daylight or cloud cover.

If a meteoroid is small enough, it will radiate away its kinetic energy as it slows down and not melt or ablate. The size of meteoroid that might survive intact depends on its entry speed, angle, composition, and its area-to-mass ratio: The mass of a meteoroid will govern how much kinetic energy it possesses, and hence how much heat is to be dissipated, while its surface governs the total area from which that heat can be radiated or conducted away. Objects with large area-to-mass ratios cool more easily, because the area scales as the square of the size, but the mass varies as the cube.

Particles smaller than about 0.1 millimeters in size (100 microns) are likely to be decelerated gradually in the atmosphere without melting or ablating. They gradually settle out, eventually reaching the surface. Such particles (typically 10 to 100 microns in size) have been collected in the stratosphere by using sticky plates attached to the wings of very high-flying aircraft. Down at the surface, such tiny particles (termed *micrometeorites*) are common, and a finger swept along a shelf that had not been dusted for a week or two would be virtually certain to collect a few micrometeorites along with the other detritus. Although such particles were first recognized late in the nineteenth century, when magnets dragged along the floor of the Pacific Ocean were found to be covered in tiny iron spherules when they were pulled up (that is, nickel-iron micrometeorites that had melted but not evaporated when zipping into the atmosphere), it is the particles collected at high altitude that provide our most pristine source of interplanetary dust. These tend to have loose, fluffy structures, which may reflect the fact that only the refractory materials are left; volatile materials that might have filled the interstitial gaps would have been lost as the particle heated on atmospheric entry.

The physical processes involved in meteor ablation were first properly understood in the 1950s when Ernst Öpik formulated the necessary theoretical framework and correlated it with the many meteor observations he had made over the years. As the space age dawned, these studies became of great practical importance, because bringing a spacecraft back through the atmosphere to the surface of the Earth meant that the problems of ablation had to be overcome. If a spacecraft attempting reentry arrives at too shallow an angle, it will bounce off of the top of the atmosphere like a flat stone skipped across the surface of a lake; if it comes in too steeply, it will burn up, incinerating the astronauts. There is a narrow band of entry angles, however, at which the vessel will encounter sufficient atmospheric mass as to be slowed gradually and to cool at the same time, given that it has enough insulation and a shield that gradually burns off—an ablation shield. The point is that it is not easy to bring home an object the size of an *Apollo* command module or the Space Shuttle, even though they are hollow and therefore have much lower masses (and consequently much lower kinetic energies) than solid meteoroids of the same size. The spacecraft also arrive at speeds of only about 8 km/sec (whereas the meteoroids come crashing in at 11 to 73 km/sec), again making their energies much lower.

Over the years, our understanding of the physics involved in the entry of meteoroids of sizes up to a few meters has improved with every new observation of a fireball. For example, the relationship between their brightness and the height that they reach before fizzling out tells us something about how their mass and speed affect the ablation process, and study of the spectra they emit tell us about their compositions and the temperatures that they attain. Really big meteoroids of sizes reaching into the regime where we would term them *asteroids*, however, enter the atmosphere so infrequently that we seldom have a chance to observe one. Because of this, we need to turn to theory and then try to fit that theory to the few observations that we do have in hand.

As anyone reading this book so far will have gathered, by far the best example we have of a large body entering the atmosphere is the explosion on June 30, 1908, over the Tungus River (Tunguska) region of Siberia. Having referred to this event several times already, it would perhaps be useful to give some account here of the circumstances. On the last evening of June and the first few nights of July 1908, the sky over Europe was noticed to be uncommonly bright. For example, *The New York Times* reported that in London "the northern sky at midnight became light blue as if the clouds were breaking, and the clouds were touched with pink in so marked a fashion that police headquarters was rung up by several people who believed that a big fire was raging in the north of London." There are reports of the game of cricket (which requires excellent light) being played after midnight in London, and just to the north, a lady in the town of Huntingdon wrote to *The Times* (of London) to remark that "It was possible to read large print indoors, and the clock in my room was quite light, as if it had been day." Those white nights have never been properly explained; it is not clear whether they were due to aurorae being induced by the energetic explosion far to the east, high-altitude dust deposited by the projectile scattering sunlight, or extraordinary noctilucent clouds (water crystals suspended at 40- to 70-kilometers altitudes) induced by the event.

The atmosphere was also the conduit by which the shock wave from the blast propagated from the impact site and encircled the planet, with sensitive barographs in Cambridge (England) and elsewhere recording transient pressure changes. Closer to the epicenter, the sound was audible more than 600 kilometers away, and wall-hangings rocked and china rattled and fell, useful indicators of the power of the concussion. Even closer to the blast, Siberian herdsmen were knocked from their feet, and on the Trans-Siberian Railroad the driver halted his locomotive. The barograph records were later used to provide reasonably accurate estimates of the energy released in the explosion, with values in the 10- to 20-Mt region being favored.

This explosion occurred a little after 7:17 A.M., local time, so the inhabitants near ground zero were starting their daily round of chores. Twenty years later, a farmer was to recall his experiences of that summer morning:

> When I sat down to have my breakfast beside my plough, I heard sudden bangs, as if from gunfire. My horse fell to its knees. From the north side above the forest a flame shot up. Then I saw the fir forest had been bent over by the wind, and I thought of a hurricane. I seized hold of my plough with both hands. . . . The wind was so strong that it carried soil from the surface of the ground.

The reason that it was 20 years after the event that his words were recorded was that it was not until 1927 that the first scientific expedition probed the region to see what might have happened, which is not surprising, given the inaccessibility of the area and the unrest in Russia leading up to the revolution in 1917.[2] The expedition was led by Leonid Kulik, who dedicated many years to his obsession with the great explosion. Kulik had expected to find a giant meteorite and crater, but there was sign of neither; nor did he find even small meteorite fragments, no matter how hard he looked, although later researchers have successfully retrieved tiny samples of melted rock and metal from the soil.

What Kulik did find was a vast expanse of fallen forest, with the trees pointing radially away from the epicenter, pushed flat by the tremendous wind generated. At the center of this array was an astounding discovery: a grove of trees that had been killed but left standing, clearly from where the downward blast had provided no transverse force to knock them over. One thing was common among the trees near the center, flattened or not, and that was that they had been charred, but generally not *burned*. It has only been in recent years that this has been properly understood: The luminous intensity from the detonation 6 to 10 kilometers above was enough to char and ignite the forest, but the following blast wave snuffed the fire out almost as soon as it had started. Further out, where the wind was less forceful, many trees were incinerated, as were reindeer, tepees, and other artifacts of the nomadic inhabitants of the region. With Kulik's expedition, the diligent scientific study of Tunguska had begun, a study that is yet to be completed. We still have much to clarify.

Although the Tunguska event is the best-known and most energetic such episode, there have been others. Siberia was also the target for an impact in 1947, when an iron projectile detonated in the atmosphere and scattered meteorites over a wide area near the region known as Sikhote-Alin, producing more than a hundred craters larger than 1 meter in size, the biggest being 14 meters across. This event was rather less energetic than the one at Tunguska, but it served both to revive interest in the 1908 event and to demonstrate that the composition of the impactor is important with regard to whether large solid bodies reach the ground. At the close of Chapter 5, I also mentioned a crater-forming impact in Estonia in 1937. In 1990, a 5-meter crater was produced near the town of Sterlitamak in eastern Russia, just to the west of the Urals, by a 1-meter iron meteorite that smashed into the ground on May 17. Lest the capitalist reader fondly imagine that such bolts from the skies might be divine messages directed at the Soviet Union (as it then was), a few large atmospheric detonations elsewhere might be listed; for example, that over Revelstoke (British Columbia) in 1965, over Kincardine (Ontario) in 1966, College (Alaska) in 1969, and the Grand Tetons/Montana in 1972. The latter would have exploded if it had approached at a steeper angle instead of bouncing off of the atmosphere and never getting deeper than 58 kilometers' altitude, despite being observed on an atmospheric track 1,500 kilometers long. Estimates of the energies of these meteoroids and asteroids range from a few kilotons up to about a megaton, possibly even higher for the Montana fireball; we are not talking firecrackers here. I guess that even this list will leave the inhabitants of the main population centers of the United States feeling pretty secure, although the Peekskill (New York) fireball of October 1992—which damaged a car with its meteorite and was observed by thousands of people because it arrived on a Friday evening during well-attended high-school football games—should have given them pause.

Although one might readily accept that rocks several meters across totally burn up as they enter the atmosphere, it is hard to believe the suggestion that a small mountain, the dimensions of a city block, would simply blow up and leave practically no trace of its constituent material. That was the serious scientific suggestion for Tunguska, leaving flying saucers, Black Holes, antimatter, and other bizarre suggestions aside. The post-World War Two era of test detonations of nuclear weapons with suitable energies had allowed the power of the Tunguska explosion to be calibrated, zeroing in on the real energy rather than the previous wild estimates of between 1 and 1,000 Mt. The area of forest that was blasted flat allowed a reasonable determination to be made, although there had been much regeneration by the time the first scientific expedition reached Tunguska 20 years later. The best indicators of the energy, however, came from the air and seismic wave records from as far away as England.[3] Having bench marks from nuclear tests, it was possible to make a much better estimate of the energy released at Tunguska, at values around 10 to 20 Mt. For a rocky body like a small asteroid, this would indicate a size of around 50 to 60 meters, depending on its arrival speed. Was it believable that such a large rock could obliterate itself in the atmosphere, never reaching an altitude even as low as the top of Mount Everest?

Because this phenomenal explosion occurred in their own backyard, it is not surprising that (former) Soviet scientists have led the way in studies of the Tunguska event, even though their efforts have been largely ignored by many in the West. In the past few years, several expeditions to the region have been mounted, with Gennadij Andreev of the Tomsk University taking a lead in bringing together researchers from many disparate fields: astronomers, geologists, meteoriticists, botanists, biologists, and so on. The expeditions, with the involvement of scientists from several countries, have led to an improved understanding not only of the blast damage, but also of the after effects. For example, studies have been made of whether the mutation rates of plants and insects in the blast region have been affected by the catastrophe. As we have seen earlier, such studies are of special interest because it appears that much larger impacts—the globally catastrophic events—have been responsible, at least in part, for the twists and turns in the evolution of life on this planet.

It is the phenomenon of the detonation of large bodies in the atmosphere, however, that is of main interest to us here, rather than the Tunguska projectile's effects on the target region below. The first solid simulation of the Tunguska event seems to have been that carried out in Russia by Vladimir Korobejnikov and colleagues,[4] using a complicated (of necessity) model of the dynamical interaction of a hypervelocity solid object with the fluid atmosphere, which led to an assessment of the physical nature of the Tunguska object. One of the fundamental characteristics of the Tunguska event, which it has in common with many very bright fireballs but a smaller fraction of fainter meteors, is the fact that a distinct explosion occurred; that is, this was not a case of a gradual ablation, the body losing mass until little was left, as is the case for smaller particles. If all solid bodies entering the atmosphere behaved as smaller meteoroids do, the only difference would be that the larger bodies would take longer to ablate, and so would penetrate deeper before fizzling out. Most of those larger than 5 to10 meters in size would reach the ground to form craters. This is *not* what actually happens, however.

During the entry of a large (say, 50-meter) body into the atmosphere, a very small fraction of its kinetic energy is dissipated through the drag applied by the upper layers of the atmosphere and the ablation (boiling off) of material from its surface. As it gets deeper, however, the atmospheric pressure climbs steeply, and in a single second, the density of air through which the object must pass climbs by a factor of 10 to 20, depending on the steepness of the entry angle and the speed. To a hypervelocity object, this is like hitting a brick wall, and so it breaks up into myriad smaller pieces, all moving along basically the same direction as the original monolith. Just before the disruption, the single projectile was reasonably effective in burrowing through its gaseous surrounds, but now it has fragmented into a large number of separate lumps, each of which is subject to deceleration and ablation on its own. The result is an explosion, because the kinetic energy of the whole is surrendered quickly, the drag forces on the individual fragments being much higher than the relatively aerodynamic monolith just prior to its disruption. The impression that this most brings to mind is a scene repeated in countless Hollywood movies, where the bandits are just about to make their escape from the cops in a small airplane with a suitcase full of money. But by some mischance (or malevolence on the part of the scriptwriter), the suitcase drops from the airplane. As it falls through the air, it bursts open and the thousands of bills it contains spill out, billowing in the wind to form a huge cloud that slowly falls to Earth, scattering over a wide area. The point is that before the suitcase burst open, it was falling with all its contents very quickly, but after its lock failed, the bills spread themselves over a wide area and their fall was rapidly arrested. It is just this sort of thing that happens when a large meteoroid fragments—bang!

Clearly, the height at which such an explosive fragmentation takes place will depend on the usual litany of parameters: the entry speed and angle, projectile size, shape, composition, and so on. For example, a 100-meter cometary object (which we would expect to have a predominantly icy composition and a mean density about half that of water) would be expected to blow up at a high altitude, whereas a solid nickel-iron asteroid 40 meters across (and thus having the same mass as the 100-meter comet) might reach the ground largely intact. You need a bullet-proof vest to protect yourself from an assassin's bullet, whereas an umbrella could do a good job of saving you from a snowball, even if it were traveling at high speed. An interesting application of such ideas was developed by Fred Whipple. He pointed out that to shield spacecraft from impacts by small meteoroids and dust, all that is needed is a thin shield of metal held some centimeters from the main spacecraft skin. What then happens is that any meteoroid hitting the shield is fragmented, and although it passes through, it will have been split into many fragments which spread laterally. These may then be absorbed by the spacecraft skin without causing significant damage, whereas the complete meteoroid will have punched a hole. This "bumper shield" concept was used successfully by the European Space Agency on its *Giotto* spacecraft, which flew within 600 kilometers of the nucleus of Comet Halley in 1986, being hit by a huge number of meteoroids as it did so.

Christopher Chyba and Kevin Zahnle (of NASA–Goddard and NASA–Ames Research Centers respectively), working with Paul Thomas (of the University of Wisconsin at Eau Claire), recognized that the solution of the long-standing puzzle of Tunguska—why such a sudden, tremendous explosion was seen at a height of about 10 kilometers—could be realized if they were to develop a realistic model of the atmospheric entry of small bodies. They recognized that for small meteoroids, which we see ablating as meteors, the physics is relatively well understood, and for very large bodies, such as 1-kilometer and larger asteroids and comets, the atmosphere is essentially nonexistent from the perspective of the projectile; that is, it suffers little or no deceleration or ablation (compared to its huge mass) prior to striking the surface. The reason for this, in a technical sense, is that the shock wave produced within the body as it meets the denser part of the atmosphere has insufficient time to cross that body before it reaches the ground.

Considerations of this point revolve around the speed of sound. A shock wave propagates through a medium at the speed of sound within that medium, which depends on a number of factors, including its density and composition. The speed of sound in air at sea level is about 330 meters per second, which is why you see a batter at the baseball stadium hit the ball before you hear the crack of the bat on the ball—the sound takes maybe a second to reach you, whereas the light from the bat and ball arrives at your eye practically instantaneously. Within a solid or a liquid, the speed of sound will be higher by a factor of perhaps 5 or 10, but within a loose agglomerate of material—such as we might imagine an asteroid or comet to be, with interstitial holes and faults that will not transmit the shock—the speed of transmission of a shock wave will be lower. A rough guess for its velocity might be the same as the speed of sound in air. If we now assume for a moment that a large body—say 1 kilometer in size—can reach a height of 10 kilometers before such a shock is produced by the increasing density of the atmosphere, then this shock wave starts to propagate about half a second before it hits the ground. Because our guess at the shock speed means that it would take a few seconds to cross the whole body, the projectile hits the ground before it starts to disintegrate and explode under the influence of the shock. An example of this sort of phenomenon, which is too often seen in the media, is a train crash. The locomotive at the front and the first few carriages will be badly damaged if the train runs into another train or a stationary barrier at full steam, but the trailing carriages will be relatively unscathed and the passengers therein shaken but largely unhurt. This is because the shock wave from the impact passes poorly through the length of the train due to the weak coupling between the carriages. If the train consisted of one huge carriage, however, the shock wave would propagate more easily, resulting in the passengers at the rear getting a worse shaking up than they would otherwise receive.

Thus if we consider the whole mass spectrum of solid particles, the very smallest are gently decelerated and reach the ground intact as micrometeorites; the ones between the sizes of dust and good-sized boulders tend to burn up completely and little or no solid material is left; while those that are very large (above about 100 to 200 meters) meet the ground at virtually the same speed as that which they had in space, which means very quickly indeed. It is those in the 10- to 100-meter size range that are poorly understood in terms of what happens to them as they enter the atmosphere at hypersonic velocities. It was these that Chyba and colleagues decided to investigate.

The model that they used for the projectile was simple—too simple to give a definitive result, but good enough to generate a realistic idea of the phenomena involved. It was simplistic in that they performed calculations for objects that were assumed to be cylindrical in shape moving along the direction of its axis of symmetry, and assumed to be monoliths with single compositions (that is, not inhomogeneous mixtures of, say, rock and metal). The most important factor affecting the heights at which detonations might occur is the physical strength of the projectile, and although we can get a good idea of this for stone or metallic bodies (from the meteorites found on Earth), we are essentially ignorant of the strength of cometary material. Some idea of the resilience of comets can be derived from the fact that we have observed several comets to have broken asunder during close passages by the Sun or Jupiter, and calculations on that basis render extremely low tensile strengths. In fact, even for stony meteorites, the tensile strengths measured are less than the aerodynamic stresses experienced by an interplanetary projectile of even the lowest possible speed (about 11 to 12 km/sec), as it approaches the 10-kilometers altitude.

In reality, we expect projectiles from space to be irregular in shape and to be spinning, so that the cylindrical, nonrotating model used by Chyba, Zahnle, and Thomas cannot be the final word, but their results are important and hopefully dispel forever the Black Hole, antimatter, and flying saucer ideas. They initially assumed that five different impactor types were feasible, corresponding to iron, stone, and carbonaceous[5] asteroids, and short-and long-period comets. Appropriate densities were assumed for each. The asteroids were all presumed to arrive at 15 km/sec, while the comets were differentiated solely on the basis of the short-period comet arriving at 25 km/sec and the long-period one at 50 km/sec. All were taken to have an entry angle of 45°, which we have already seen is the most likely case, and to have sizes appropriate to a kinetic energy equivalent to 15 Mt of TNT (fixing the diameters to range from 40 meters for the long-period comet up to 68 meters for the carbonaceous asteroid).

Chyba, Zahnle, and Thomas found that their model predicted the detonation of such projectiles at altitudes of 29 kilometers for the long-period comet, 23 kilometers for the short-period comet, 14 kilometers for the carbonaceous body, and 9 kilometers for the stoney one; the iron asteroid was found to reach the ground intact. This seems to support the idea that the Tunguska explosion, at a height of between 6 and 10 kilometers, was due to a stoney asteroid, the most common type. They then experimented further with the stoney projectiles, letting them come in at a variety of different incidence angles, and found that the altitude of the predicted detonation varied between 6 and 15 kilometers for trajectories between the vertical and the horizontal, respectively, again adding weight to the hypothesis that the Tunguska object was a stoney asteroid that blew up at close to 10 kilometers' altitude.

This all seems to be in good agreement with the observed *height* of the detonation, but Chyba, Thomas, and Zahnle ignored other important evidence indicating both its actual *entry angle* and its *speed*. It happens that the Tunguska object arrived at the time of the peak of one of the most intense annual daytime meteor showers, as was noted by, for example, Arthur C. Clarke some years back (see bibliography). But there is more to link the Tunguska object to the shower than that, as the Slovak astronomer, Lubor Kresák, suggested in 1978. Kresák studied the various eyewitness reports of the arrival direction of the object and found that they placed its apparent direction of origin—what meteor watchers call its *radiant*—slightly to the west of the Sun in the constellation Taurus. That coincides with the radiant of the daytime meteor shower, determined by using suitable radars. Kresák also pointed out that the meteors are known to be linked to Comet Encke (as discussed in Chapter 8), so it seems likely that the Tunguska projectile was a fragment of that comet. This would then define both the entry angle and the speed of the Tunguska object as it plummeted into the atmosphere: They would be identical with the parameters of that meteor shower.

The speed deduced in this way is about 33 km/sec, however, far higher than Chyba and colleagues would allow from the results of their modeling, especially if one also presumed that, being a fragment of a comet, the projectile must have been of low density. Perhaps this is why they ignored Kresák's work. From another perspective, one could try to determine the chance that an unrelated object would arrive from the direction of that shower radiant while the shower was near its peak. We cannot even begin to make a scientific estimate of that probability, but it is certainly very small; that is, I view it as being a poor piece of science to reject the observational information that one has available merely because the crude model that you have constructed does not agree with it. In fact, the results of Korobejnikov and colleagues support a putative link with Comet Encke, which was another good reason for Chyba and colleagues to ignore that work. It just didn't fit their modeling.

Can the physical approach of Chyba, Zahnle, and Thomas be reconciled with Tunguska's being genetically linked to Comet Encke? First, let us lay to rest the idea that the Tunguska object must have been of low density if it had a cometary origin. There are numerous observed Earth-crossing asteroids that are dynamically associated with Comet Encke, which itself appeared asteroidal prior to the late eighteenth century. Thus, although many comets might well be low-density objects with meager tensile strengths, it does not follow that P/Encke and the Tunguska projectile must be. They could well be characterized as basically rocky in nature.

Now, what about the model of Chyba, Zahnle, and Thomas? There are a number of objections to this model, in terms of what we *do* know and what we do *not* know about comets and asteroids. One of the assumptions made is that the projectile was homogeneous, which is contraindicated by various observations (for example, Comet Halley having a dark nucleus with only a small fraction of the surface actively evaporating, the dark majority presumably being coated with silicates and heavy organics). If comets were homogeneous, how could they split? A splitting requires a weak line of fracture, implying inhomogeneity. Chyba and colleagues found that most of the energy of their model comets was lost through ablation rather than rapid deceleration in a detonation, with the converse being true for the asteroids. Thus a projectile having a heterogeneous, highly differentiated structure—which is not excluded by observations of comets—could behave similarly to the observed Tunguska object. For example, a 100-meter icy lump with a large stoney core and a dark crust produced from differential material loss while in space (that is, the volatiles evaporating to leave a silicate/heavy organic crust, in accord with our knowledge of Comet Halley) might behave in a way appropriate to fit the observations: The crust would soon be stripped away, the ice layer would ablate in the way in which Chyba and colleagues predict with their model (at high altitude), and then the stoney core would detonate at around 10 kilometers. The objection that a stoney asteroid would not penetrate to 10 kilometers if it were traveling at over 30 km/sec is therefore dismissed, because an icy coat may allow it to penetrate deeply before exploding. An oft-quoted aphorism is that "truth is stranger than fiction." One should not forget that the model of Chyba, Zahnle, and Thomas was a piece of fiction, even if it was an important step forward.

Chyba and colleagues rejected an entry speed of 30 km/sec or more for the Tunguska object on the basis that, with their simple model, the detonation would occur higher than the maximal 10 kilometers altitude actually observed. In an earlier chaper, we discussed tsunami generation by small asteroid explosions on the basis of the work by Jack Hills and Patrick Goda. Their work also entails modeling of asteroid entry into the atmosphere. They developed a rather more sophisticated model than that of Chyba and colleagues, producing results that do not contradict the hypothesis that the Tunguska object was a large member of that meteor shower. Hills and Goda found that a hard stone projectile entering at 30 km/sec would indeed fragment or detonate at an altitude of about 10 kilometers. Perhaps the suggestions of Kresák and Clarke are not wrong after all, and the Tunguska object *was* a chunk from Comet Encke.

In their extensive paper, Hills and Goda presented the results of their calculations for a wide range of input parameters, covering the physical and dynamical variations expected for asteroids (solid iron, hard stone, and soft stone) and comets (rather fluffier, lower-density bodies). Their computations, taking into account fragmentation of the incoming objects, rendered many useful plots showing the altitudes at which the certain events of interest would occur: for example, how the energy of the projectile varies with height, how bright it would appear to an observer on the ground, the speed and mass of any fragment that might reach the ground, and so on. The effect on the Earth's surface was also considered: For objects detonating in the atmosphere, the area devastated by the blast wave was computed, while for projectiles reaching ground level at high speed, the resultant earthquake magnitude (on the Richter scale) was determined.

Whether the projectile reaches the ground largely intact or not, a huge tsunami would be produced; Hills and Goda calculated the height of the resultant tsunami at a range of 1,000 kilometers from ground zero below the detonation. In fact, the energy deposited either into the atmosphere (in the case of a detonation) or the surface (when the projectile survives entry) produces what is known as a deep-water wave, which propagates outward across an ocean. The tsunami itself is the wave that breaks on the shore of a landmass, this typically being about 10 times the height of the deep-water wave, although the amplification in the wave height may be by a factor of 100 to 120 in some cases, depending on the topography of the coastal region. An asteroidal object about 100 meters in radius produces a deep-water wave a few meters high at a distance of 1,000 kilometers from the explosion, and therefore one expects a tsunami perhaps 100 meters high at this range. As a rule of thumb, the tsunami height is about the same as the size of the projectile at 1,000 kilometers' range, and the height drops off as the inverse of the range. In a large ocean, a tsunami can have a very long range indeed. We saw that a tsunami produced by an earthquake in Chile in 1960 killed dozens of people in Japan, more than 17,000 kilometers away. From this point of view, one of the most dangerous places to live is on the fringe of a large ocean. In particular, the Pacific Ocean poses more than 30% of the entire target area of the Earth, so that a good fraction of incoming asteroids and comets will produce tsunamis that can affect the cities around its edge. The implications of this for the citizens of Los Angeles, San Francisco, Tokyo, Shanghai, Sydney, and so on are worrying; Shin Yabushita (Kyoto University, Japan) has calculated that there is at least a 1% chance that all of the cities around the Pacific rim will be obliterated by an asteroid-induced tsunami within the next century.

There is one drawback, however, that I would like to mention with regard to the excellent modeling of Hills and Goda. In fact, my criticism is not in connection with their model as such, with which I can find little fault, but rather with their input parameters. Earlier we discussed the impact speed of an object coming from interplanetary space and meeting the top of the atmosphere, which can be anywhere between 11 km/sec (the terrestrial escape velocity) and 73 km/sec (corresponding to an object on an orbit that is retrograde, going around the Sun in the opposite direction to the Earth, but highly eccentric and with perihelion at 1 AU, so that it meets us head-on). The perceived wisdom is that asteroids mostly strike the Earth at less than 20 km/sec, whereas all the high-velocity impacts are due to comets. Thus Hills and Goda only performed computations for their objects of asteroidal composition (iron, hard and soft stone) having arrival speeds of between 11 and 30 km/sec, but for their cometary objects having speeds from 20 to 70 km/sec.

I can argue against the 30-km/sec limit on a number of grounds. First, if one considers the Earth-crossing asteroids discovered so far, more than 10% of them have orbits which dictate that any impact would be at greater than 30 km/sec; there is also growing evidence that the presently known sample of such asteroids may have lower eccentricities and inclinations than the norm, implying that higher impact speeds than 30 km/sec may be more prevalent. This makes a big difference to the kinetic energies of the incoming objects. Second, if one looks at the speeds determined for meteors—that is, small objects that *are* actually hitting us, as opposed to large objects seen in space which *may* hit us—one finds that a very large fraction (more than 50% in some meteor surveys) have arrival speeds in excess of 30 km/sec, and yet have sub-Jovian "asteroidal" orbits. The full range of behavior of asteroids meeting the atmosphere was therefore not covered, which is a shame, considering the immense usefulness of the work of Hills and Goda apart from this omission.

If the Earth is so well "protected" from small asteroids and comets, in the sense that our atmosphere prevents most interplanetary projectiles smaller than 100 meters in size from reaching the ground intact, what must be the situation for Venus, with its dense atmosphere and surface pressure one hundred times that of our own planet? A clue is given by recent spacecraft data. Images from the radar on board the *Magellan* spacecraft have shown that Venus has a surface that is heavily cratered: Obviously some asteroids and comets do get through. Few craters smaller than about 8 kilometers in diameter appear in the images, however. This deficiency indicates that impactors smaller than about 1 kilometer in size do not penetrate to the surface. This has been explained by Kevin Zahnle with the help of the same sort of model as he applied with Chyba and Thomas to the Tunguska event. Not many projectiles reach the surface of Venus, but because the temperature there is near 500°C, the pressure is about 100 bars, the atmospheric gas is mainly carbon dioxide, and the clouds above are made of sulfuric acid droplets, the Earth remains a nicer place to live.

10

Great Efforts, Great Ignorance

In Chapter 2, mention was made of the fact that the first Earth-crossing asteroid was discovered in 1932, and we have also noted that the first dedicated search for such bodies began in the 1970s. Now, we will discuss the various search programs currently in operation, what they have found, and what yet awaits discovery.

From the amount of media coverage that the asteroid/comet impact hazard receives, one might believe that there are whole armies of astronomers scouring the skies, hunting these things down. That might enable you to sleep better at night, thinking that while you snooze, your welfare is being looked after, but in reality nothing could be further from the truth. As David Morrison first pointed out, the total number of people engaged in this sort of work worldwide is less than the staff of an average McDonald's restaurant. As a consequence, we can cover only a small fraction of the sky, and indeed much of the equipment we use is outmoded, suboptimal, or not always available, being used by other astronomers for much of the time.

The three Earth-crossing asteroids that were found in the 1930s were picked up because, at that time, wide-field telescopes were becoming available, with reasonably efficient photographic plates. Previously, the available photographic materials had been very poor and, in any case, were used with the conventional type of telescope, which covered only a very small area of the sky. For most astronomical research, what is required is a telescope that collects light from only one star or galaxy, rather than forming an image of many objects. In the 1930s, the first Schmidt-type telescopes became available. These telescopes, named for the German astronomer who invented their basic optical system in the 1920s, cover wide swathes of the sky. In fact, some people call them Schmidt *cameras* rather than *telescopes*, because they are basically very large cameras, even though they use a mixture of lenses and mirrors instead of the pure lens-optics of a conventional camera.

The problem is that for a normal telescope using a large paraboloidal primary mirror to collect the light,[1] and then a hyperboloidal secondary to focus it onto the detector, as one strays far from the optic axis, the images become aberrated; that is, they have flaws such that the light from a star, say, is spread out and the resolution is lost. This is not a problem for most astronomical purposes because, as mentioned, the idea is to collect light from just one star or galaxy that will be placed on-axis, perhaps then feeding the light through a spectrograph so that various physical parameters of the object may be determined. However, the angular width of the sky that may be covered by such a telescope, and still produce acceptably low aberrations, is about half a degree at the most, which is similar to the angular diameter of the Moon. But now consider the fact that the Moon covers only about one part in 100,000 of the hemisphere visible from any location, meaning that at least 100,000 photographs would be necessary if one wanted to map the entire sky—and twice that to cover both the northern and southern hemispheres.

Schmidt came up with a design that could photograph a much larger area: depending on the actual implementation, perhaps 100 to 200 times the lunar area, meaning that only 500 to 1,000 photographs would be needed for each hemisphere. This is a much more manageable quantity and makes complete photographic surveys of the heavens feasible.

The essence of Schmidt's design is that by placing a large, specially shaped lens at the top of the telescope tube, the incoming light could be distorted prior to meeting the primary mirror. The amount of distortion is calculated to exactly compensate for the aberrations produced off-axis, making it possible to produce clear photographs over a wide area. Even with this convoluted corrector lens, it was still not possible to produce a flat focal field, with stars and galaxies instead being focused onto a curved surface. Photographic plates—emulsions on glass plate backings still in use today, although superior film has been increasingly substituted during the last decade—had to be bent while they were being exposed. Even with plates only a millimeter thick, this bending obviously led to plenty of breakages. The plates are also large (36 centimeters, or 14 inches, on a side, for the large Schmidts), making handling quite arduous.

The apertures of the biggest Schmidt telescopes were again limited by how large a corrector lens could be manufactured. Two Schmidts in particular have been noteworthy for their productivity, providing complete sky coverage in surveys that are now being repeated with a view to assisting the pointing of the Hubble Space Telescope. These are the large Palomar Schmidt (in California) and the United Kingdom Schmidt (in Australia). These are twins, in design if not in age; the former started work in the late 1940s, while the latter entered operations a quarter-century later. Each has a primary mirror 2 meters in diameter, but an aperture that is limited by their corrector lenses, which are 1.2 meters (4 feet) across. It is usual to quote this aperture when describing a Schmidt. There is one larger Schmidt telescope, at Tautenburg (in Germany), but it is not located at as good an astronomical site. There are several other Schmidts with apertures of more than 1 meter, mostly situated in the northern hemisphere. The European Southern Observatory operates a 1.05-meter Schmidt, high in the Chilean Andes, which has done important work.

The big Schmidts in California and Australia have been used to photograph the entire sky several times over. The reasons for this repeated coverage are many, but two in particular should be mentioned. One is that, by using different filters, it is possible to distinguish between stars (or other phenomena) of different colors. One wavelength band might be useful for one purpose, while another band is better for some other type of search.[2] The second reason is that, although it is not obvious, the stars move relative to each other. By photographing the sky with at least a decade between exposures, it is possible to determine how fast the stars are moving, knowledge of this speed being useful for a number of reasons. For example, studies of the apparent motions of nearby stars tell us about how the Sun is moving about the galaxy and so on. Astronomers guiding the Hubble Space Telescope need to know where a star might be at the time of each observation rather than some years before. If one knows how fast a star is moving, one can predict where it will be when the telescope is scheduled to observe it.

The first Schmidt telescopes began operation in the 1930s, one of the earliest being a 0.46-meter aperture version at Mount Palomar. As more and more wide-field photographs of the sky were taken, it was inevitable that asteroids would be detected by telescopes at Palomar and elsewhere. Until this time, almost all known asteroids were large, bright, main-belt objects, although some Mars-crossing asteroids and even one or two with perihelion distances less than 1.3 AU (called *Amor-type asteroids*) had been found. For example, 433 Eros was found in 1898, 719 Albert in 1911,[3] 1036 Ganymed in 1924, and 1627 Ivar in 1929. Despite their collective name, they are not particularly lovely: More recent work on their orbital evolutions has shown that, even though they cannot come too near the Earth at the moment, in the long term (many thousands of years), they may achieve Earth-crossing orbits, making impacts on our planet possible.

The archetype of the Amor class (1221 Amor) was found in 1932, the same year that the first Earth-crosser (1862 Apollo) was spotted. No longer could it be denied that there were large, dark bodies on orbits that could intersect the Earth. Since Newton's time, it had been thought (if any thought was given to it at all) that cosmic impacts could occur only on timescales of the order of at least tens of millions of years, because only comets were known to be on Earth-crossing orbits. The new discoveries implied that the frequency would be at least once every million years, and perhaps rather more often than that.

1862 Apollo became the archetype for this new class of asteroid (Earth-crossers in the present epoch), and we often speak of *Apollo-type asteroids* or simply *Apollos*. In 1976, a new category was initiated when 2062 Aten was found.[4] While Apollos have orbital periods of more than a year and cross the Earth's orbit near the asteroidal perihelion, Atens have periods of less than a year, coming far enough from the Sun to cross the Earth near the asteroidal aphelion. It is a matter of some debate whether there might be asteroids interior to the Earth's orbit, perhaps crossing only Venus (Cytherean asteroids) or Mercury (Hermian or Vulcanian asteroids). Such intermundane bodies would be very difficult to discover and have certainly eluded us thus far. The record for small orbits was set in December 1994 when an Aten (1994 XL_1), with an orbit, on average, only two-thirds the size of that of the Earth, was discovered; it was spotted as it came to aphelion at 1.02 AU from the Sun.

Four years later, on the heels of 1862 Apollo, 2101 Adonis was discovered, and the following year, 1937 UB Hermes. The latter caused a great deal of excitement, leading to its being given a name (officially or not), despite the fact that its orbit was not well enough determined to allow it to be numbered. The reason for the excitement was that it passed only 670,000 kilometers from the Earth, less than twice the distance of the Moon, and really hammered home the fact that these things pose a significant danger to us. It was observed for less than a week, but because it was tracked by telescopes both in Germany and in South Africa, the triangulation provided by these separated sites made it possible to determine its geocentric distance quite accurately. Nevertheless, it is a "lost" (but unnumbered) asteroid because the time span of observations was inadequate to allow a future prediction of its ephemeris; the situation was exacerbated by the closeness of its approach, under which circumstance the gravitational perturbations of its path induced by the Earth and the Moon are very significant. Hermes remains the asteroid that many astronomers would most like to see rediscovered in order to remove the anomaly of being named but not numbered. For example, it annoys me because it sticks out like a sore thumb in my orbit catalogue. Such accidental rediscoveries do occur from time to time; for example, the near–Amor 1927 TC was found and then lost in that year, but 65 years later, we rediscovered it at the Anglo-Australian Observatory, meaning that it is now secure and numbered. One problem, though, is whether the original discoverer is still around to stake a claim on naming it!

After the 1930s, there was a hiatus in near-Earth asteroid discoveries, due partially to World War Two. However, especially with the fresh availability of the large Palomar Schmidt, which was then coming into operation, the late 1940s saw a boom in identifications: 2101 Oljato in

1947, 1685 Toro and 1863 Antinous in 1948, 1951 Lick in 1949, and 1580 Betulia in 1950. Most notable of all was 1566 Icarus, found in 1949. This object had the smallest perihelion distance of any known asteroid (hence its naming), passing closer to the Sun than Mercury, and also an orbital period not much more than 1 year, which at the time was considered highly unusual. Its record small perihelion was broken only when 3200 Phaethon (another appropriate name) was discovered in 1983 by using data from the Infra-Red Astronomy Satellite.

No mention has been given so far of how asteroids are identified on photographic plates. We have seen that *asteroid* means "starlike"; so, how are these objects differentiated from stars? The answer lies in the way in which a Schmidt telescope is operated. It was noted earlier that Schmidts are like huge cameras, but unlike a simple 35-millimeter camera or a pocket instamatic, the exposure durations are not mere fractions of a second. The aim is to detect very faint stars and galaxies (and other celestial phenomena), and for that reason, a very large aperture is needed so that as much light as possible is collected.[5] In addition, as much detail as is feasible is wanted, and this is limited by the atmosphere to angular sizes of about 0.5 to 5 arcseconds, depending on weather conditions. The photographic emulsion used therefore must be very fine-grained, having grain sizes that cover only a fraction of an arcsecond in terms of the amount of the sky being focused on each grain. As any photographer knows, a fine-grained emulsion means a slow film. The emulsions used on Schmidt telescopes might be rated at something like ASA 1 or 2, nothing like the ASA 64, 100, or 400 of roll films. This, of course, means that long exposures are necessary, and although the shutter might be open for only 20 minutes in some circumstances, more often the durations of exposures are two or three hours. As one can imagine, it is immensely frustrating when a planned three-hour exposure is started under clear skies and clouds roll in two thirds of the way to completion.

Schmidt plate/film exposures are of long duration, then, and as they proceed, the telescope is very accurately tracked across the sky to compensate for the Earth's spin, keeping pace with the stars. In fact, in latter years, to make this tracking as precise as possible, auxiliary telescopes strapped to the side of the large Schmidts have been focused on only one bright star in the field. Via computer control, these telescopes keep that star on the central cross-hairs of that telescope, and hence the main Schmidt is pointed at exactly the same area. This is especially important when the field being photographed is far from the zenith, because then refraction of light in the atmosphere means that the rate at which the star field moves across the sky is not quite identical to the spin rate of the Earth.

On the finished plate, the distant celestial objects (stars and galaxies) are distinct images, because they have not moved relative to each other during the exposure. Any object within the solar system will have shifted, however. For example, Jupiter will have moved about 140,000 kilometers (twice its radius) during a three-hour exposure, meaning that it would be smeared out; during the same period, the Earth would have moved about 320,000 kilometers, and the two motions would combine to make Jupiter appear as a broad streak between pinpoint images of the stars.

The same sort of reasoning applies to asteroids. The asteroids in the main belt, due to the combined effect of their own motions and that of the Earth, produce trailed images of length equivalent to about two arcminutes (one thirtieth of a degree) in a three-hour exposure, corresponding to a trail about 2 millimeters in length on the photographic plate. On any plate taken in the general direction of the ecliptic, perhaps 100 to 200 such trails might be found, all roughly aligned because they move in broadly the same direction, leading to the idea that they are "the vermin of the skies." There are so many asteroid trails that astronomers interested in stars and galaxies find the asteroids a nuisance.

If one searches such plates using binocular microscopes (which are necessary for such short trails, these often being very faint), occasionally an anomalous trail is found. This may be longer than usual, at a peculiar angle (the main-belt asteroids having predominantly right ascension motion), or (paradoxically) shorter than usual. Such trails, in a few cases, turn out to be asteroids quite close by the Earth (by which anything from the lunar distance out to an astronomical unit or more is implied). Observations over the following few weeks may then allow an asteroid's orbit to be determined with enough accuracy that it would be possible to pick it up again the following year or whenever (due to the vagaries of celestial mechanics) it next appears bright enough in the nighttime sky.

It was this basic mechanism that led to the discovery of a handful of Earth-crossing asteroids between the 1930s and the late 1960s, although because the Schmidt plates obtained around the world were not being scoured routinely for suspicious trails—remember that very few astronomers are interested in asteroids and have their hands and minds full of their own research—it is clear that the vast majority of near-Earth asteroid detections were going unnoticed on the plates.

In earlier chapters, the role that Gene Shoemaker played in the recognition of impact craters on the Earth and Moon was discussed. Coming from a geological background, but nevertheless with an astronomical bent, as the Apollo lunar landings were ending, Shoemaker wanted to branch out into another area, but one that was directly related to the Apollo endeavors. Instead of concentrating solely on the scars formed when asteroids and comets strike the Earth, he wanted to study these things while they are still in space. There were several good reasons for this, involving questions raised by the Apollo results—indeed, questions that in some instances still require a satisfactory answer. For example, from the terrestrial cratering record, it is not possible to say much about the flux of objects hitting the Earth or their size distribution. We have already discussed the multifarious reasons for this: weathering and geological turnover causing the loss of craters, the majority of impacts occurring in the oceans, preferential loss of smaller craters, shielding of the Earth from smaller impactors by the atmosphere. Paradoxically, the best place to learn about how often and how hard our planet gets clobbered is the Moon. Conversely, although the lunar craters preserve information on their relative ages (because overlying craters indicate which came first), their absolute ages are not easily determinable because, unlike the Earth, datable stratigraphy due to sedimentary deposition and other factors does not occur on a body like the Moon, which has no active geological or hydrological cycles. On top of that, we only have rock samples from a handful of locations on the near-side of the Moon, and those (due to considerations of astronaut safety) are not necessarily the most interesting, selenologically speaking.

With the results from the various lunar satellites (such as *Orbiter* and *Surveyor*, plus the Soviet probes) and the Apollo program in hand, Shoemaker and his colleagues were able to determine such things as the areal density on the lunar surface of craters of different sizes, and hence the relative numbers of impactors of different sizes (subject to various assumptions, such as the impact speeds involved). By identifying the oldest surfaces on the Moon, and counting craters per unit area, it was possible to estimate the average cratering rate per million years (again subject to a variety of assumptions, such as this rate being time-invariant—of course, we now know this not to be the case, given the evidence of periodic cratering waves on the Earth). The next question was, "How do the cratering rates derived from the lunar record compare with the numbers estimated from observations of asteroids and comets on Earth-crossing orbits?"

If one knows the orbits of the objects involved, it is possible to calculate an impact probability on the Earth for each, for example, using some simple equations derived by Ernst Öpik (whose other work was mentioned earlier) in 1951. Again, there are various assumptions that need to be made, and it is known that these assumptions are contravened by many Earth-crossing bodies. Nevertheless, a reasonable estimate of the impact probabilities may be derived in this way. Then, if the total population of possible impactors is known, an overall collision rate for the Earth (or Moon) may be derived. The problem was that, although one could be fairly sure that 50% or more of the short-period comets larger than 1 kilometer in size had been discovered, and the trans-Earth flux of large parabolic comets might also be inferred from the data in hand, back in the early 1970s only a handful of Earth-crossing asteroids was known. That this could only be a tiny fraction of the total number was obvious from the fact that whenever such a body was found on a Schmidt plate, almost invariably it turned out to be an unknown member of the population rather than one that had previously been observed.

With fewer than 20 Apollos in hand—crude estimates suggesting that there must be 1,000 or more larger than 1 kilometer in size—it was not possible to derive a reasonably secure value for their mean impact probability on the Earth. Shoemaker needed a larger sample, and the best way to obtain that sample was by doing the searching himself. In 1972, he started what is known as the Planet-Crossing Asteroid Survey (PCAS), using the small (and ancient) 0.46-meter Schmidt telescope at Palomar. His collaborator on that program, as she had been in the earlier cratering work, was Eleanor Helin.

If they had merely taken long exposures with the Schmidt, as was described earlier, the amount of sky that they could cover would be limited, because one might obtain only three or four long-exposure shots in a night. For asteroid detection, long exposures are not required, per se, because as the asteroid moves, its image is focused onto another part of the emulsion—the exposed density is not gradually built up, as is the case for a faint star or galaxy. Short exposures may therefore be used, although long time gaps are needed in order to identify the motion. The answer is to take pairs of five- or ten-minute exposure photographs separated by about an hour and see what objects have moved between times. This is what Shoemaker and Helin set about doing.

Normally, in astronomy, the way in which the comparison of two such photographs is done is to "blink" the pair: The two are mounted in an optical device that allows images of the two to be alternately displayed in an eyepiece. Then all extra–solar system objects appear stationary, not having moved between the two shots, whereas the point-like images of an asteroid or planet will seem to jump backward and forward between the two positions it occupied during the exposures. For example, this is the way that Pluto was discovered in 1930. Shoemaker and Helin, however, developed a method they found to be more efficacious.

Instead of blinking the two photographs to make the two appear alternately to both eyes, they built a device that brought an image of one shot to the left eye and the other shot to the right eye. The device is similar to that used by cartographers in studying stereo photographs of the ground taken from aircraft, the slightly different view angles producing a three-dimensional effect. In fact, this is not really three-dimensional, but *stereo*, viewing. In the case of the astronomical photographs, what is perceived is that all of the stars and galaxies appear to be in the same plane, whereas any object that has moved (an asteroid or other solar system body) seems to pop up out of that plane or sink beneath it due to the action of the eye-brain combination, which evolution equipped to interpret twin images in terms of depth.

This invention made the inspection of the pairs of photographs much more efficient, an important factor when staff is limited and there are millions of stellar images on the films, from which a few asteroids need to be distinguished. Finding a needle in a haystack can be straightforward, even if time consuming, as long as someone has invented a suitable search technique (like the use of a magnet).

Even using this artifice, the work was still slow, with many sets of photographs needing to be laboriously inspected for each new discovery. When a potential near-Earth asteroid was identified, there would then be a scramble to obtain follow-up observations so that its orbit might be secured before bad weather or the full Moon (which hinders observations due to making the sky bright) set in. Through the 1970s, Shoemaker and Helin more than doubled the number of known Earth-crossing asteroids, a remarkable achievement given the small number that had been found in the four decades since the first had been turned up by accident.

Advances in science often follow from advances in technology, and in the early 1980s, a new photographic emulsion became available—Kodak's Tech Pan film. This led to another acceleration in the discovery rate, the efficiency of the PCAS program having been boosted by the implementation of this film. In the meantime, however, Shoemaker had widened his interests to include objects of other types. In 1982, he began the Palomar Asteroid and Comet Survey (PACS), leaving Helin in charge of the similarly acronymed PCAS. Although the aims of PACS includes near-Earth asteroids, in addition, there has been some focus on Trojan-type asteroids,[6] comets, and distant solar system objects. Gene Shoemaker was joined in PACS by his wife Carolyn Shoemaker, and, later, by David Levy. Since 1982, the PACS and PCAS teams have used the Palomar 0.46-meter Schmidt for about a week each a month, making use of the dark time far from full Moon. In 1993, Gene Shoemaker formally retired from the U.S. Geological Survey, and although he maintains an office at the USGS Branch of Astrogeology in Flagstaff, Arizona, he is now beginning a new asteroid search program in collaboration with Ted Bowell at the Lowell Observatory, which is also in Flagstaff (discussed later).

A third near-Earth asteroid search program making use of photographic observations with Schmidt telescopes is my own at the Anglo-Australian Observatory (AAO). We utilize photographs taken with the U.K. Schmidt Telescope (UKST), which since 1988 has been administered as part of the AAO (previously, it was an outstation of the Royal Observatory, Edinburgh). In 1989, having seen what was being achieved by the PCAS and PACS teams, I realized that there was much that could be done with the plates and films being routinely exposed with the UKST, and so in collaboration with Rob McNaught and Ken Russell,

I obtained funding from the Australian government to start a new search program, the first one covering the southern sky. This program is called the Anglo-Australian Near-Earth Asteroid Survey (AANEAS), the acronym being chosen to echo the name of Aeneas, the only Trojan hero to escape Troy when it was sacked by the Greeks.

Our modus operandi is simple. We inspect all plates and films taken with the UKST (for other purposes) for trails that may indicate an object of interest. This may sound simple, but it entails many eye-straining, back-breaking scans of plates holding images of millions of stars and galaxies, marking each asteroid trail and comparing each with the positions of known objects. We find a near-Earth asteroid on about one plate in 50 to 100, each taking between 30 and 60 minutes to scan (depending on the star density). Any positive identification means that we then obtain a follow-up photograph, using either the smaller Schmidt telescope owned by the University of Uppsala (Sweden), which is located at the same site and operated as part of the Siding Spring Observatory, or else the UKST. The Uppsala instrument is adequate if the object is bright enough, the UKST being necessary for a faint asteroid.

With astrometric positions from at least two nights, we are then able to predict the asteroid's positions for the next few weeks with sufficient accuracy that it can be picked up using a narrow-field telescope, and we use the 1-meter telescope at Siding Spring for this task. At the same time, we obtain follow-up positions for many of the discoveries of the American teams. Since AANEAS began in 1990, we have found and secured the orbits of dozens of near-Earth asteroids on plates and films taken with the UKST, compared with just five spread over the previous 17 years of operation of that telescope. This raises the question of what our state of knowledge would be now, had a rigorous search been done of the photographs obtained with other large Schmidts. There is much that could be done, but too few people to do it.

The use of photographic film, however, is becoming very much outdated. Practically all observing time on large (conventional) telescopes is now dedicated to light detection using electronic detectors called *charge-coupled devices* (CCDs). Whereas a photographic plate might register, at most, 8% to 10% of the photons of light that hit it (and usually rather less), CCDs are now available that detect 80% or more of the photons. Thus, similar results may be obtained in much shorter exposure times or by using much smaller telescopes. In addition, the images are available for immediate computer analysis rather than having to wait to chemically process the unwieldy photographs.

CCDs do have drawbacks, however. First, because their individual pixels (which may be as small as 10 microns across) are larger than the grains in photographic emulsions, in principle they do not have the same resolution as photographs; in practice, one matches the size of the image produced by the telescope to the pixel size, and also to the size of the "seeing disk" (the blurring due to atmospheric scintillation). If one has pixels covering a quarter of an arcsecond, and the seeing is half an arcsecond, then one would not gain much from having smaller pixels. Second, and of pertinence here, CCDs cannot (yet) be produced to cover the same physical area as photographic plates, and this means that they have yet to be used extensively in Schmidt telescopes. The largest CCD chips, with 4,000 or 5,000 pixels on a side and measuring perhaps 8 centimeters square, fill only a small fraction of the focal field of a large Schmidt. Even by forming a mosaic of individual CCD chips, it is possible to cover, at most, 50% of the field, because there are problems with the physical dimensions of the wires needed to read out the data. There have been experiments along these lines with the large Schmidt at the National Astronomical Observatory in Japan, but, in general, the world's Schmidt telescopes remain the major professional instruments where photographic techniques are predominantly used.

One smaller Schmidt, with which it is intended that a small CCD mosaic will be used, is at the Lowell Observatory. As mentioned earlier, Gene Shoemaker and Ted Bowell are starting a new search program, which will be called the Lowell Observatory Near-Earth Object Survey (LONEOS). The idea is that they will mount four CCD chips and come close to filling the field of the 0.41-meter Schmidt that they have available. They are also intending to replace the present corrector lens with a 0.56-meter aperture lens (the same size as the primary mirror), and in this way almost double its light grasp. If their modeling is correct, they will be able to discover near-Earth asteroids at an unprecedented rate, even though theirs is quite a modest-sized telescope.

We began by emphasizing that Schmidts are useful because they cover wide areas, this being necessary for the discovery of objects of unknown celestial location. We also noted that the asteroids in question move quickly across the sky, and as a result, they smear out their light along a trail on a photographic plate (implying that asteroids with intrinsic brightnesses similar to the faintest stars in the field will not be detected because their light is spread too far as a consequence of their motion). If we now pose the same sort of problem to asteroid detection using a CCD, we see that nothing is gained by having an exposure duration that is longer than the time it takes the asteroid to traverse one pixel: A shorter exposure results in fewer photons being detected by that pixel, while a longer exposure merely means that the adjacent pixel receives the photons. If we imagine a telescope with pixels 0.5 arcseconds wide, in the case of a typical near-Earth asteroid moving at an angular speed of one degree a day, that asteroid crosses a pixel in 12 seconds of time, meaning that nothing (in terms of number of photons from the asteroid counted by that pixel) is gained by having an exposure time of more than 12 seconds.

We might therefore recommend that we take a series of 12-second CCD exposures, and then compare them to see what has moved. There is another problem with CCDs, however, and that is the length of time it takes to electronically read them out so that the data are stored in the computer. Few large CCDs will read out in under two minutes, which would mean that there must be a two-minute gap between each 12-second data collection period; to look at it another way, the telescope would actually be observing for less than 10% of the time, and in a ten-hour observing night, only one hour of data collection would occur—clearly, an unsatisfactory state of affairs. For observations of a faint field of galaxies, it would be just fine to take a one-hour exposure and then wait through the two minutes of dead time for the CCD to read out, but for short exposures (as dictated by the asteroid motion), the efficiency would be very low.

Soon after CCDs started to become available, Tom Gehrels of the University of Arizona saw a brilliant way around this. He asked why the CCD could not be read out at the same time the exposure is being made. In fact, CCDs are read out (and thus the long read-out time) by shifting each column of pixels one by one across the CCD chip; first the column nearest the read-out terminals is taken, then the one next to it (which in the meantime has moved across), and so on, until all have been taken. As each column shifts across (say, from right to left), the electrons stored in each pixel remain in the same row (position up or down). Instead of waiting for this process to be completed, Gehrels decided to incorporate it into the observing process. He saw that if he switched off the telescope drive that compensates for the spin of the Earth, the image would drift across the CCD chip; but if he read out the CCD columns at *exactly* the same rate as this drift, the data fed into the computer memory would be a strip scan across the sky.

Gehrels' implementation of this idea is called the *Spacewatch Telescope*. This is a 0.91-meter aperture, narrow-field instrument[7] at Kitt Peak near Tucson, Arizona. A CCD with 2,048 pixels in each dimension is used; that is, 2,048 columns and 2,048 rows, about 4 million pixels in all. This covers a little more than the diameter of the Moon on the sky, around 0.6 degrees square. With the right ascension drive switched off, the image takes about two and a half minutes (150 seconds) to cross the field, and the pixels are each a little more than one arcsecond wide. This means that a near-Earth asteroid moving at one degree per day would cross a pixel in about 26 seconds. In 150 seconds, it passes through six or so pixels, forming a small streak; this might be regarded as being suboptimal from the perspective of detection efficiency, but from the contrary view, the streaks may aid the automatic recognition of a moving object.

The Spacewatch team, which includes Gehrels, Jim Scotti, David Rabinowitz, and Robert Jedicke, makes use of that telescope for 18 nights each month (each lunation); over the bright-of-Moon period, the instrument is used for other purposes. Each night, they select their search areas and then scan the region in question three times. Typically, they scan for 20 minutes, then move back to the start position and scan again, and finally scan a third time. Each scan would then cover about 3 square degrees in total. In a good night, they might therefore cover about 20 to 25 square degrees, about half the area of a photographic plate taken with the UKST or the large Palomar Schmidt. Their limiting magnitude is about 21, however, whereas the asteroidal magnitude limit for the Schmidts is about 19, implying that Spacewatch can detect asteroids more than six times fainter and thus patrol a larger volume of space (because they can detect objects of the same intrinsic brightness but further from the Earth).

Why are three scans needed? One might imagine that the computer could be programmed to search for moving objects using only two scans. The answer lies in another area of astronomy: cosmic rays. These are elementary particles from the Sun and elsewhere in the galaxy which are continually flooding the Earth. Most are absorbed in the atmosphere, but some punch through and reach the ground. Indeed, countless billions have passed through your body since you began reading this chapter. Although we do not notice these cosmic rays, some may have an effect on a CCD. It is possible that such a particle will hit a particular pixel of the CCD and cause it to saturate; that is, it seems as if the pixel has detected a large flux of photons, whereas, in fact, it was a single cosmic ray that struck it. This is a familiar, and unavoidable, problem to all astronomers using CCDs. Its implication for the Spacewatch team was that a 20-minute scan effectively contains about 200 million pixels, and thousands of these will have been affected by cosmic rays. This is only a problem when a pixel in one scan has been struck, and by mischance a nearby (in the context of the image of the sky) pixel in the next scan has also been struck. In this situation, which arises fairly often, the computer software might erroneously interpret the two cosmic rays as being evidence of an asteroid that has moved between scans. By taking a third scan, this problem is circumvented, because the likelihood of getting three nearby cosmic ray strikes is very small. Remember the Bellman's affirmation in *The Hunting of the Snark*: "What I tell you three times is true."

Using these techniques, Spacewatch began routine observations in 1989. Since then, about 40% of all near-Earth asteroid discoveries, along with several other objects of interest with regard to the history of impacts on the Earth[8] have been identified by that team. Although the number of large (say, greater than 0.5-kilometers) asteroids found by Spacewatch is similar to the scores of the other three routine search programs (PCAS, PACS and AANEAS), Spacewatch is, in addition, uniquely able to detect and track small (less than 100-meter) asteroids as they pass close by the Earth (and so have very high angular velocities). Thus Spacewatch has revolutionized our knowledge of the flux of these objects. Previously, our sole information source had been small (less than 1 kilometer diameter) lunar craters, because most 100-meter asteroids do not reach the ground and so do not produce terrestrial craters. We therefore could not predict with any confidence how often events like the Tunguska explosion might occur. Clearly, it is not once per year, but is it once per 50 years or once per millennium? The Spacewatch data indicate that a rate closer to the higher frequency is likely (that is, every century or less; "worse", if you like). In fact, those data have been interpreted as implying that the terrestrial influx of 10- to 100- meter objects is one to two orders of magnitude higher than previously believed (which is in line with an extrapolation from the fireball observations that produce values for the influx of 1-meter meteoroids), although there is still some argument about this because the flux calculated is critically dependent on the individual sizes, and those are estimated using assumed albedos that may or may not be reasonably close to the truth.

Because near-misses of the Earth are of interest to people, it might serve to reel these off. After Hermes in 1937, as described earlier, there was no observed closer passage until 1989 when 1989 FC (now called 4581 Asclepius) missed us by 650,000 kilometers. That object is a few hundred meters in size and was discovered by the PACS group. Because of the unique Spacewatch capabilities, from 1989 that team has excelled in recognizing near-misses. In early 1991, they found 1991 BA, which missed by 170,000 kilometers (less than half the distance to the Moon). 1993 KA_2 came a shade closer in May of that year, and 1994 ES_1 missed by a similar margin in March 1994. As I do the final editing of this book, the record holder is 1994XM_1, which missed us by 104,000 kilometers in December 1994. These are all objects of 5- to 20-meter sizes. These comprise only a small fraction of the total that passes closer than the Moon, however, with more than 99% undoubtedly escaping unseen. This ratio may improve somewhat in the near future because Spacewatch under Gehrels is planning to implement a larger (1.8-meter) telescope; but it will certainly still be a matter of spotting only a minor fraction of the Earth-missers.

Even if Spacewatch could scan 20 square degrees every night, this is less than a thousandth of the celestial hemisphere. If they scanned a *different* region each night, each month they could still cover less than 2% of the sky. The Spacewatch data are invaluable in inferring what the near-Earth flux might be, but by no means are they currently catching anything but a minor fraction of the small asteroids whizzing by, and no one else is seeing any at all. It is therefore not surprising that the 10-meter or so asteroid that blew up over a largely vacant area of the western Pacific on February 1, 1994, producing an explosion equivalent to at least ten times that of the Hiroshima bomb (and possibly rather more), was not seen prior to impact. Surveillance satellites registered it as the brightest such explosion that they have picked up so far. Despite the efforts of numerous scientists in this area of study to make the military aware that such detonations do occur naturally, it appears that the U.S. President was awakened because the Pentagon thought that this incident might be a hostile nuclear explosion. Asteroids and comets are dangerous in more ways than one: We need to be aware of them lest an ill-timed and ill-placed natural event mistakenly provokes a nuclear war.

Although most near-Earth asteroid discoveries in the past decade or so have been made as a result of the specific search programs mentioned here, in fact, from time to time, other individuals have found a previously unknown object. Two people in particular should be mentioned: Jean Mueller, who works at the large Palomar Schmidt, and Christian Pollas of the Observatoire de la Cote d'Azure (OCA) near Nice in southern France. As time allows, they scour plates taken with the Schmidts at those locations, looking mainly for supernovae (exploded stars) but also occasionally turning up an anomalous asteroid trail. In fact, a new near-Earth asteroid search program is planned at OCA, involving Alain Maury of the Nice Observatory and Gerhard Hahn of the German Space Agency. As in LONEOS, a mosaic of CCD chips will be used.

It has been pointed out that the discovery rate of near-Earth asteroids has increased immensely in recent years. Although the first Apollo was identified in 1932, the known population of these has doubled since 1990. Although many of these are small objects, it is true that we have now found many more of the 1-kilometer-plus asteroids that threaten a global catastrophe than we had catalogued only 5 years ago. However, we still know of only a small fraction of the total population of such objects: few scientists involved in this area believe that we have to date discovered much more than 5% of that total. Although none of the known asteroids is going to hit the Earth in the foreseeable future (the next century or two), this is not a particularly comforting fact, because if there *were* an asteroid due to strike home soon, then there is a greater than 95% chance that we would not have found it yet.

The reader might now take comfort with the notion that, because our discovery rate has increased, surely that 95% will soon be reeled in? Again, this would be a delusion. If we continue as we are at present, it would take perhaps 500 years to complete the search for all the Apollos larger than 1 kilometer, and longer for the Atens. Thus if one has "our number" on it for the year 2025, we would most likely not find it ahead of time. Like the small asteroid explosion in the Pacific in February 1994 or the Tunguska explosion, the first we would know about it would be when it is too late. This might not be a problem if the asteroid were only 10 meters in size, or even 100 meters; but what if it were a kilometer or more across? Under such circumstances, one thing is certain: You would not hear about it on the radio or TV or read about it in the newspapers. You would hear and feel its arrival no matter where you lived, and you would have to try to live with its consequences.

11

So Began Project Spaceguard

In the late 1980s, as various members of the scientific community realized that huge impacts on the Earth must occur with a worrying frequency, another phenomenon was occurring: The media were also starting to give coverage to the story. In a situation like this, it is not simply one event or one news item that produces a change in the public opinion, but rather the sum of many stories spread over a long period of time. Many things undoubtedly contributed—from the Alvarez team's promotion of the "dinosaur extinction by impact" idea through to reports of the "near-miss" by 1989 FC mentioned in the previous chapter. Change in public opinion is a matter of gradual sensitization, and over the past few years, the concept that we might be at risk from cosmic impacts has seeped into the collective consciousness of most Western countries.

In no nation was this change in perception more extensive than in the United States, with its dynamic media ready to respond to any story of interest to its consumers, especially if a mega-disaster is involved. As a result of this and the political system in the United States (with elected representatives needing to be seen to respond to their electors' wishes), some action was soon perceived to be desirable on the part of Congress. The usual chain of events is that scientists or engineers within NASA decide what they would like to do (such as send a new spaceprobe to Mars); they then argue their case in front of the NASA administration against others who support alternate projects (such as building an x-ray satellite to study the cosmos); and then the NASA hierarchy makes its decision and takes its request to those who control the purse strings. In the case of asteroid impacts, however, quite the opposite happened. Although there was no great interest (compared to the other powerful lobbies) within NASA to consider the near-Earth asteroid problem, the agency received from Congress a direction to investigate the matter.

The House of Representatives, in its NASA Multiyear Authorization Act of 1990, stipulated the following:

> The Committee believes that it is imperative that the detection rate of Earth-orbit–crossing asteroids must be increased substantially, and that the means to destroy or alter the orbits of asteroids when they threaten collision should be defined and agreed upon internationally.
>
> The chances of the Earth being struck by a large asteroid are extremely small, but since the consequences of such a collision are extremely large, the Committee believes it is only prudent to assess the nature of the threat and prepare to deal with it. We have the technology to detect such asteroids and to prevent their collision with the Earth.
>
> The Committee therefore directs that NASA undertake two workshop studies. The first would define a program for dramatically increasing the detection rate of Earth-orbit-crossing asteroids; this study would address the costs, schedule, technology, and equipment required for precise definition of the orbits of such bodies. The second study would define systems and technologies to alter the orbits of such asteroids or to destroy them if they should pose a danger to life on Earth. The Committee recommends international participation in these studies and suggests that they be conducted within a year of the passage of this legislation.

Clearly the politicians had started to take the matter seriously, at least insofar as attempting to obtain an expert opinion.

Two committees (or workshops) were necessary, then. The first, chaired by David Morrison, might be termed the *Detection Committee*, whose task it was to write a brief that basically covered the *astronomical* aspects of the problem: How do you search these objects out and then determine their orbits with sufficient precision such that any potential impact in the foreseeable future might be identified? The second committee, which was initially chaired by Jurgen Rahe and John Rather of NASA HQ (later joined by Greg Canavan and Johndale Solem of Los Alamos National Laboratory in New Mexico), covered the *space* aspects of the problem and might be labeled the *Interception Committee*. Of the 24 members of the Detection Committee, I had the honor to be one of six members from outside of the United States; the Intercept Committee had more than 90 members and was barely international in that I was the only foreign member, although there were a few other members (for example, Ted Bowell and Brian Marsden) who were *born* elsewhere. The focus of this chapter is the recommendation of the Detection Committee; the work of the Interception Committee is described in the next chapter.

It is clear that the Spacewatch program has paved the way for any future search strategy. The program recommended by the Detection Committee was based broadly on the experiences of the Spacewatch team. The steps for deciding what was required are as follows.

First, one must decide what image scale is desirable on the CCD chips to be used. Although many good astronomical sites have seeing (the degree of atmospheric scintillation) that is better than 0.5 arcseconds on many nights, an average over the whole year is unlikely to be better than one arcsecond. Because we are in the business, in this specific application, of getting as much light as possible onto one pixel, this means that the pixels used should cover around one arcsecond of the sky. For a 2,048-pixel-square CCD chip, this means that a section of the sky about 0.57 degrees across is covered. In terms of area, this is less than one part in 60,000 of the hemisphere visible at any time.

The next consideration is the size of the telescope aperture required. In previous chapters, the threshold impactor size at which a global catastrophe might be induced was discussed, that threshold apparently being somewhere in the range of 1 to 2 kilometers. To be on the safe side, a limit of 1 kilometer was adopted by the Detection Committee; this may then be translated into an apparent brightness (for an assumed albedo) at some certain distance. That is, one can model in a sophisticated computer program how faint the minimum brightness of the asteroids can be such that the program would result in the detection of at least 99% of them down to that set size, within a reasonable time. That reasonable time was defined as 25 years, and the modeling was carried out by Ted Bowell in conjunction with Karri Muinonen of the University of Helsinki, Finland; Muinonen happened to be working at that time with Bowell at the Lowell Observatory. The resultant limiting magnitude (brightness) was equivalent to a stellar magnitude of 22, and that in turn implied that telescopes with apertures of 2 meters or larger are needed.

Given that most CCD chips have pixels that are about 25 microns across, these two limitations (pixels covering one arcsecond; aperture, 2 meters) then lead to the necessary focal length of the telescope—5.2 meters. In terms of the f-ratios used in photography, which may be familiar to many readers, this calls for an f/2.6 optical system, and this defines the curvature of the mirror needed. If the decision were made to make larger telescopes (and there are good arguments for this, as discussed next), the value would be f/2.1 for a 2.5-meter aperture, or f/1.7 for a 3-meter aperture.

Why go to larger telescopes? Two reasons spring to mind. First, although a 1-kilometer asteroid impact would likely cause a severe global upset of the terrestrial environment, an impactor half this size is not to be sneezed at. Impacts by half-kilometer objects occur about four times as often; so why not try to find those, too?

Second, the modeling is based on objects in orbits like those occupied by the presently known near-Earth asteroids, which do not stray more than about 4 AU from the Sun. Our present perception is that these pose the majority of the impact hazard. Compared with asteroids, comets are reasonably easy to detect, so short-period comets would be expected to be found quickly in the planned program. Parabolic comets, however, approach from the depths of space, and we have no way to predict their arrival. While far from the Sun they remain dark, brightening somewhat as their more volatile constituents begin to evaporate in the outer solar system; they do not start to get very bright until they reach 3 AU from the Sun, whereupon their load of water ice becomes unstable, at least on the exposed surfaces. This water then starts to evaporate and the comet swiftly brightens. There is therefore another reason for choosing to build as large a search telescope as is possible—finding parabolic comets while they are still far away. Some researchers believe that such bodies pose only 2% of the overall impact hazard, while others claim 25% or more. It may well be that the answer to this problem will be received only by carrying out the search program now under consideration.

We return now to the design of the search telescopes. We saw in the last chapter that, during the scan time of the Spacewatch telescope, a near-Earth object with an angular velocity of 1 degree per day covers about six pixels (or 6 arcseconds), and there is little profit (with regard to detectability) in allowing this to occur. All that is required is that the light fall on one pixel during a scan. In the case of Spacewatch, the team is limited by the rate at which the Earth spins, and that defines the rate at which the image of the sky crosses the CCD and hence the number of pixels in which the object is detected (only one being the desired result for the sake of efficiency). The answer is quite simple: Rotate the telescope to conduct a scan across the sky faster than the Earth spins (the sidereal rotation rate). For a 2-meter aperture, the necessary drive rate is six times the sidereal rate, as might be expected (to get the image in one pixel only, instead of six). For larger telescopes, the required drive rate is higher.

The next question concerns the amount of the sky that can be covered in this way. For the set-up in question (covering a swath about 0.6 degrees wide), if this scans the sky at six times the rate at which the Earth rotates, then the area of coverage is about 0.85 square degrees per minute. Multiple scans are required, however—three, from the experience of Spacewatch. Allowing for dead time between scans as the telescope is repositioned, the total new area scanned per minute would average to about 0.25 square degrees. This would lead to object identification; more sets of scans would be required on subsequent nights in order to obtain more astrometric positions for the objects, thus allowing their orbits to be determined. A good, clear astronomical site might provide, at most, 80 or 90 hours of suitable observing time each month, and allowing for follow-up observations, around half of this (40 hours per month) would be available for searching. At 0.25 square degrees per minute, this means that about 600 square degrees would be covered each month by the single CCD-telescope pairing under consideration.

The modeling of Bowell and Muinonen showed, however that the search telescopes would need to cover about 6,000 square degrees per month between them in order to complete the program in 25 years.

This is ten times that mentioned earlier, which would suggest that ten telescopes would be needed. Great economy might be gained by having several CCDs in each telescope, however. The most obvious solution is to have a mosaic of four CCDs on each telescope, implying a field about 1.2 degrees wide; although that is more than the usual limit obtainable by reflecting telescopes without auxiliary optics, it is feasible for instruments as fast (f/2.6) as discussed here. If there were six telescopes in all, they could, in principle, cover more than 14,000 square degrees per month (that is, 2.4 times the nominal 6,000-square-degree requirement), but bad weather, among other conditions, would reduce this area. Six seems a sensible number of telescopes, then.

The question that now springs to mind is the location of these telescopes. An equable distribution in temperate or near-tropical latitudes is needed. There is no doubt where the telescopes would go, at least for the southern hemisphere. They would need to be placed in South America (Chile), southern Africa, and Australia. In the north, there are a number of possible locations, but one might assume that the existing excellent astronomical sites in Hawaii and the Canary Islands would be used, along with at least one other site (preferably more) at a different longitude.

What part of the sky would these telescopes search? The optimal search area, as is clear from Bowell and Muinonen's modeling, is a wide region centered on opposition (the spot exactly opposite the Sun) spread up to 60° north and south of the ecliptic and a similar angle in longitude (and thus the 6,000 square degrees). To return to the question of comets, although this search region is optimal for near-Earth asteroids and short-period comets, Brian Marsden has shown that such a search would miss many of the incoming parabolic comets. In particular, those coming from the polar directions would not be picked up in advance. One way to avoid this would be to spread the search region (which would make the detection of the periodic objects less efficient) or else employ two larger (more than 4-meter aperture) telescopes in searches of the outlying regions. Larger apertures are required to detect the comets when they are still beyond Jupiter. Only a pair would be required, one in each hemisphere, because an annual scan of these regions would be sufficient, given that it takes some years for a parabolic comet to get to the Earth from Jupiter.

The necessary search plan was defined, then, subject to the aforementioned considerations. The main plan calls for six 2-meter (preferably larger) specialized telescopes spread around the globe and acting in concert. This describes the baseline program. If it is decided that parabolic comets also pose a substantial hazard, then a pair of 4-meter telescopes of a similar design would be needed.

The obvious question, next, is, "How much would it cost to carry out this detection program?" The answer is about $300 million spread over 25 years (that is, about $12 million a year; at most, 1% of the calculated liability, according to the calculations in Chapter 1). *Not* to carry it out makes no sense, economically speaking, being equivalent to turning down reputable and secure automobile insurance offered to you at $1 a year. On top of that, the benefits from the search program would continue to accrue after the 25 years was up: Having identified the objects, it is straightforward to keep tabs on them in the future.

But how was the $300 million figure determined at? This would be the cost of the baseline six, 2-meter-telescope network. First, an admission: the Detection Committee looked at what it had been instructed to accomplish by the Congressional mandate and then asked, "How much might we reasonably expect to be made available?" The answer to this question was determined to be "about as much as the cost of a small space mission," which is around $300 million, from the experiences of NASA. Personally, I argued against this position: My view is that a spectrum of possibilities should have been given in the report, describing what could be done in at least three categories of response:

1. Increase present funding by a few million dollars per year, and we can double or triple the present discovery rate, finding all 1-kilometer-plus objects within a couple of centuries or so;
2. Fund the baseline six-telescope network at $300 million; or
3. Spend billions of dollars and solve the problem in short order (although there is a limit imposed by celestial mechanics), finding most, if not all, objects down to a few hundred meters in size.

So that it won't be imagined that I am deciding what I think the American taxpayer should spend money on, I hasten to point out that (quite rightly) it was not proposed that the United States fund the entire program; rather, the United States might build and operate two of the six telescopes, with other nations responsible for the others. Certainly, I see no good reason why Australia should not be responsible for one of the telescopes, especially because it would need to be sited there.

The defined figure was $300 million, then, and by happenstance that is also the total funding needed to complete a search for all objects down to the threshold causing a global catastrophe. The breakdown is as follows: Each telescope would cost about $6 million to build, if manufactured as a batch of six (because then there are savings from duplication). The CCD arrays and so on would cost about $1 million per telescope, and similarly, the computer equipment and peripherals needed to cope with the huge data rate would come in at about $1 million per site. This totals $8 million per telescope, or $48 million for the set. The provision of a single operations center from which the network would be controlled, at about $2 million, brings the total for hardware to $50 million. Each installation would require about 11 staff, including astronomers, telescope observers, hardware and software engineers, administrative back-up, and technical support. The overall costs of this personnel, along with the staff of the central nexus, would cost about $10 million per year, or $250 million over the planned 25-year program. All added up, this comes to $300 million quite conveniently.

One could now argue about where this $300 million should come from. As I have already noted, it would be entirely appropriate that the funding be spread among several countries, perhaps being run through the United Nations or some similar body. Most people, however, would imagine that it should be funded from some sort of science budget. Let me disabuse you of that notion. What is being talked about here is a defense project, not a scientific one. Certainly, scientific understanding would be derived should a program like that outlined proceed, but that would be a byproduct. The purposes of science would be fulfilled should 25%, or even 10%, of the potential impactors be studied. There is no need to study every kangaroo in Australia to understand that species' feeding habits, nor every armadillo in Texas to discover that they produce quadruplets of the same sex. We would know quite enough about asteroids and comets to explain their role in the general scheme of things if we studied 20% of the total population, but that would not tell us whether or not one of the other 80% was due to land in Ohio in the year 2013.

The sort of program just discussed is one based on present-day scientific and technological capabilities. In the past, there have been many science fiction stories of varying merit and credibility based upon the basic idea—asteroid menaces humankind; search it out by some means, then divert or destroy it. Toward the fiction end of the scale (as opposed to fact) was the 1979 movie *Meteor*, even its title being a misnomer. While the Detection Committee was drafting its report, David Morrison pointed out the existence of such movies as *Meteor* and such books as *Lucifer's Hammer*, by Larry Niven and Jerry Pournelle. I sent Morrison a message indicating that a search program based broadly along the lines of what we were considering had been described by my friend Arthur C. Clarke[1] in the first chapter of his novel *Rendezvous with Rama* (1973). Clarke described an asteroid a few hundred meters in size entering the atmosphere above Europe in the year A.D. 2077 and impacting in northern Italy, discussing the number of people killed and the damage done. Clarke then went on to write the following:

> After the initial shock, mankind reacted with a determination and a unity that no earlier age could have shown. Such a disaster, it was realized, might not occur again for a thousand years—but it might occur tomorrow. And the next time, the consequences could be even worse.
>
> Very well; *there would be no next time.*
>
> A hundred years earlier a much poorer world, with far feebler resources, had squandered its wealth attempting to destroy weapons launched, suicidally, by mankind against itself. The effort had never been successful, but the skills acquired then had not been forgotten. Now they could be used for a far nobler purpose, and on an infinitely vaster stage. No meteorite large enough to cause catastrophe would ever again be allowed to breach the defenses of Earth.
>
> So began Project SPACEGUARD.

And so, indeed, our planned search program for offending asteroids and comets got its name: Spaceguard. Not only did the name say it all, but also nicely echoed the name of the Spacewatch program on which it is based.[2]

Clarke's novel also brings up the question of radar. Most readers will surely be aware of the various uses of radar techniques in looking for unknown objects (like straying airplanes or incoming missiles) and so have probably been asking themselves, "Why use optical telescopes for searching for asteroids? Why not use radar?" Indeed, Clarke envisioned using radar in his invention of Spaceguard. There are several reasons, however, why radar cannot be used, at least with our present technological capabilities. But who knows what we might be able to devise in a few decades?

The first reason that radar cannot be used is the doppler-shift problem. Any echo from a moving object has its frequency shifted. This is the way, for example, that the police can determine an automobile's speed by using a small radar gun. For Earth-based objects, however, the speeds are quite small, the frequency shifts therefore are small, and generally the radar receiver bandwidth can cope with them. For example, a radar operating at wavelengths around 10 centimeters has a frequency of 3 Gigahertz (GHz) and might have a receiver bandwidth of 100 kilohertz (kHz). Even an intercontinental ballistic missile traveling at 5 kilometers per second produces a frequency shift of only 50 kilohertz, so that the echo would still be within the bandwidth. If one were patroling interplanetary space for unknown asteroids, however, these would have typical speeds relative to the Earth of 20 km/sec, and parabolic comets could have relative speeds of 72 km/sec, implying frequency shifts of 700 kilohertz or more, meaning that the radar receivers would need to be set up to interrogate signals from a wide variety of frequency bands, all of them susceptible to interference from various natural and artificial sources of noise.

The second reason that radar cannot be used is concerned with range. On the Earth, an airport radar might cover ranges out to 500 kilometers, while military radars searching for missiles might cover ranges out to thousands of kilometers. Radio waves move at the speed of light (300,000 kilometers per second), so even if the maximum range were as much as 3,000 kilometers, the radar could still send out pulses at a rate of 50 per second and receive all echoes prior to the next pulse being transmitted. This is not the case for interplanetary distances, however. It takes a radar pulse more than eight minutes to cover a distance equal to an astronomical unit, so if one were looking at a search parallel to that accomplished by the optical Spaceguard program, covering distances out to perhaps 4 AU, one would only be able to send out one pulse per hour! Pulse coding can circumvent this drawback to some extent, but there would still be difficulties.

Third, the energy returned in a radar echo drops off quickly with distance. The intensity of light or any other electromagnetic radiation falls off as the inverse square of the distance from its source. Thus at a range of 4 AU, the radio energy per unit area in a radar pulse is less than one part in 2 million of what it was at the distance of the Moon. When the energy is reflected or scattered by an asteroid, that asteroid acts as a new source, and the radio energy returning to the Earth again drops off as the inverse square of the distance to the asteroid. Thus the amount of returned energy depends on the inverse fourth power of the range and therefore would be tiny.

The fourth reason that use of radar would not be effective is the wide range of sky to be covered. To obtain sufficient antenna gain to discern a detectable echo, the radar beam would need to be constrained to a narrow cross-sectional area, but then a very restricted area of the celestial sphere would be covered.

Finally, although there are various sources of noise that could swamp any weak signal, as mentioned previously, in fact, the radar would effectively produce its own noise. The problem is this: Suitable wavelengths to use in such an interplanetary radar are in the centimeter-decimeter band, 10 centimeters being a nominal choice. There are reasons why wavelengths much longer (more than 1 meter) or much shorter (less than 1 millimeter) cannot be used. Recall that the objects we want to detect are asteroids and comets 1 kilometer in size or larger; even if we wanted to guard against all impactors which could reach the ground, we are still talking about 100 meters being the minimum size limit. A fundamental law of physics is that the scattering efficiency of an object depends on the inverse of its size[3] as compared with the wavelength. This means that objects from 10 centimeters to 1 meter or so in size will be the most efficient scatterers of the radio waves from a 10-centimeter radar. The problem is that there are many more of these than there are 100-meter or 1-kilometer asteroids. In a distribution of particles, such as exists in the interplanetary medium (with the number varying roughly as the inverse square of their sizes), although the majority of the *mass* is held in the larger particles (big asteroids), most of the cross-sectional *area* is held in the smaller ones. Thus the zodiacal light (discussed earlier) is due predominantly to light scattering by dust particles 10 to 100 microns in size, not asteroids, even though the mass of the asteroids is far higher than the accumulated mass of the dust. In the case of a 10-centimeter radar, looking for asteroids would be like looking for a snowman in a blizzard.

In the previous discussion, I have simplified some of the arguments, but overall it is clear that radar has no role in asteroid *searches*. But does it have any role at all? The answer is a resounding *yes*. Radar of the type described has been used to obtain echoes from a wide range of solar system objects. Starting in the early 1960s, vast, powerful radars such as those at Goldstone (in California) and Arecibo (in Puerto Rico) have been used to get echoes from, first of all, the Moon, then Venus and Mars, and later, Mercury, the moons of Jupiter, and the rings of Saturn, to list only a few. Such studies provide important information on the surface structure and composition of these objects. For example, the first maps of the surface of Venus, perpetually shrouded by dense clouds that cannot be penetrated using optical telescopes, were obtained in this way; these maps have since been superseded by radar data obtained by the *Pioneer Venus* and *Magellan* spacecraft in orbit around the planet. Evidence of subsurface permafrost on Mars was obtained in the late 1970s by Earth-based radar, re-opening the question of how the flow channels on Mars were formed and whether microbial life (at least) might exist there. A recent surprise from radar observations has been the identification of ice near the poles of Mercury; no one had expected the temperature to be cold enough so near to the Sun, but the source of that ice remains a mystery. Similarly, there is radar evidence for ice near the poles of the Moon, which may be of use to future lunar colonies.

In addition, numerous comets and asteroids have been detected by radar. The leading scientist in this endeavor has been Steve Ostro of the Jet Propulsion Laboratory in California, and because of his expertise, he was also a member of the Detection Committee. Given that one knows to within some tolerance where an asteroid will be at a certain time (so that the radar can be pointed in the precise direction and the times of the returned echoes can be predicted) and that one also knows its speed relative to the Earth (so that the radar receivers can be tuned to the appropriate doppler-shifted frequency), it is possible to detect an asteroid provided that it is coming close enough by our planet. The maximum range clearly depends on the size of the asteroid, but is typically around 0.2 AU; large comets like P/Halley have been detected at greater ranges, although part of the returned signal in such cases is due to reflections from the cloud of meteoroids surrounding the nucleus, forming part of the coma.

In a few cases, the asteroid has come close enough that it can actually be imaged through an ingenious technique; that is, a picture of the shape of the asteroid can be built up. One surprise has been that at least two asteroids (4769 Castalia and 4179 Toutatis) have each appeared to be bifurcated, both looking almost like two separate objects glued together. Others have rendered echoes with strengths that indicate they are likely to be vast nickel-iron bodies. From the perspective of an impact hazard assessment, however, the most significant factor is the determination of asteroid orbits.

A radar detection can render the asteroid's geocentric distance to within a few kilometers and its line-of-sight velocity to within a few centimeters per second. This contrasts with optical astrometric observations, which tell us only where the asteroid is in the plane of the sky to within (typically) 1,000 kilometers. In terms of improving our knowledge of the precise orbit of an asteroid, a single radar detection is equivalent to the summation of 10 years of optical tracking. If we want to be in the business of detecting asteroids and then be able to predict whether or not they will hit the Earth in the foreseeable future, it is essential that we define their orbits very accurately. For most asteroids, the optical data will very quickly show that the asteroid cannot strike the Earth soon, because its orbit crosses the ecliptic far from 1 AU. For those asteroids that are identified as having such crossing points near 1 AU, however, it would be highly desirable that their orbits be defined swiftly, making radar detections virtually essential. The provision of planetary radars could therefore be seen as an adjunct of Spaceguard as planned. The optical telescopes will identify any potential hazards, but radar would be needed to determine their orbits quickly, and therefore, whether an impact is likely or not.

Apart from Goldstone and Arecibo, there is also a radar sometimes run as a joint Russian-German exercise, with transmission of pulses from a radar in the East and reception with the large Bonn radio telescope in Germany. With some upgrading, and most especially the provision of more personnel to supplement Ostro's great efforts, these radars might conceivably do the job, although there are some drawbacks with sky accessibility. For example, one cannot wait and hope that the dangerous objects are going to pass through those declinations that you can observe.

There is one great gap in the sky coverage—the entire southern sky. The three radars mentioned are all in the northern hemisphere. There is no such device in the southern hemisphere. My own country, Australia, has great expertise in radio astronomy and radar, and I am informed that it is the only southern-hemisphere nation that has the necessary technological capability and infrastructure to build and operate such a radar. I have tried to lobby for such a beast to be built in Australia, but without success (so far). Obviously, this is not yet seen as being a problem, at least among those who control the purse strings. One must not be too glib, however, about demanding the provision of such installations. Radars of the type mentioned, with radio dishes perhaps 100 meters in diameter and exceptionally powerful transmitters, do not come cheap. The total cost of ensuring the availability of radars in the northern and southern hemispheres would be hundreds of millions of dollars, quite likely comparable to the cost of Spaceguard itself.

I entitled this chapter "So Began Project Spaceguard," a direct quote from Clarke's *Rendezvous with Rama*. Doing so may have led the reader to believe that Spaceguard has actually begun—far from it. Although the Spaceguard report was read into the Congressional record in March 1993, so far nothing significant has happened (but see the Epilogue). NASA made some extra money available, but it ended up being much less than a million dollars, which is barely enough to keep the current U.S. search programs continuing. This is not, as such, a criticism of NASA; the point is that the major injection of funding has yet to come through, although there have been some recent developments (see the Epilogue). I have tried to alert the two countries of which I am a citizen (Australia and the United Kingdom). In Australia, our program continues to exist on a shoestring budget. Despite my urging, the governments of other nations are similarly sitting back, either unaware of the problem or unwilling to do anything about it. Perhaps, as Arthur C. Clarke wrote, we will need to await a catastrophe before the governments of the world say, "There will be no next time."

12

Showdown at Los Alamos

When I was a child, there was a hit song that contained the refrain, "Catch a falling star and put it in your pocket, never let it fade away." That's a pretty good description of what we need to do to protect ourselves against impacts: Catch the falling stars with our telescopes, continue to track them (not letting them fade away), and, if one happens to be on a collision course, put it in our pocket, or at least constrain it insofar as we change its orbit to keep it away from our planetary home.

In the previous chapter, we saw that the report of the Detection Committee recommended that a search program—the Spaceguard Survey—be carried out over a period of about 25 years, as a result of which essentially all asteroids capable of wreaking a global catastrophe on the Earth would be identified and their orbits determined. The minimum size capable of causing such a catastrophe is about 1 kilometer, meaning that an estimated 2,000 to 3,000 asteroids would compose the main aim of the project, although in carrying out that survey, a larger number of smaller Earth-crossing asteroids would also be discovered. Those smaller asteroids, however, would represent a decreasing fraction of the total population at each size limit (that is, although more than 99% of the 1-kilometer-plus asteroids might be found, there might be 50% of the 0.5-kilometer ones and only a few percent of those above 100 meters).

Actually finding and tracking such bodies, however, would not, of itself, compose a defense system, because we would be showing only that the end of the world (as envisioned by the apocalypse merchants encountered in many cities) is indeed nigh, without providing the means for that conflagration to be averted. The first question, then, concerns whether humankind is technologically able to avoid such a disaster; one might note that *disaster* literally means "bad star." It seems, at first thought, that it is a mighty task to shove a 1-kilometer asteroid with a mass of more than a billion tons, traveling at more than 30 kilometers per second, out of the way of the Earth. What needs to be done?

First of all, we must recall that the Spaceguard plan was not a program to identify asteroids shortly (meaning days, weeks or months) prior to their arrival on the Earth. The idea is that as large a fraction as possible of the population of potential impactors is identified and their orbits are then followed forward with sophisticated computer programs for a period covering the "foreseeable future," which is limited to the next century or two by the onset of chaos in the orbits. Of course, this limitation is not a problem from the perspective of defending humankind against the threat because we may be fairly confident that our great-grandchildren and their descendants will be able to look after themselves. If not, they will have worse troubles than the possibilities discussed in this book.

One lesson from the media coverage and popular response surrounding the Comet Swift-Tuttle panic of late 1992 is that many people found it difficult to take seriously a possible impact more than a century into the future. Consequently, ever since that flurry of excitement, I have chosen a fictitious impact date closer to the present when discussing with the media or in public lectures the possibility of an asteroid or comet impact. One needs to emphasize that if by mischance there *is* an asteroid or comet on the way to hit Earth, it is unlikely to be due next year. Conversely, people are prone to dismiss an impact predicted for the year 2099 as being of no immediate concern. The number I usually pick (perhaps because it rolls off of the tongue rather nicely and is also a prime number) is 23. As we will see in the next several paragraphs, without my realizing it when I first started using that number, for reasons of round figures, it was quite a fortuitous choice.

The question that I first want to address is, "How hard must we push the errant asteroid in order to make it miss?" (I will later discuss *how* we might perform the pushing and shoving.) For the sake of argument, let us assume that we are able to provide an impulse to the asteroid such that we are able to change its velocity by just 1 centimeter per second. To repeat that, we are talking about a velocity change of only a centimeter per second compared with a speed in space measured in tens of kilometers per second. This does not seem like very much, but would it be enough?

The worst-case scenario would have the asteroid due to come in vertically onto the Earth's surface, meaning that its extrapolated path would pass right through the center of the Earth. Under such circumstances, the amount that it would need to be diverted is equal to the radius of the Earth plus a hundred kilometers or so for the thickness of the atmosphere, remembering that meteor ablation occurs at about that height—let's say 6,500 kilometers for the deflection required.

Now, we apply our push to the asteroid 23 years ahead of the predicted impact, and the impulse is applied sideways (that is, at right angles to its direction of motion). This change in velocity propagates forward for all of that time; that is, it continues to move sideways at 1 centimeter per second more than it would have done, so that its orbit and path are slightly different from those that it would otherwise have followed (and which would result in an impact). There are about 30 million seconds in a year, so the sideways motion amounts to 30 million centimeters, or 300 kilometers, in each year. In 23 years, the total sideways motion is 6,900 kilometers, meaning that the asteroid passes 400 kilometers above the atmosphere, and humankind and our fellow inhabitants of the Earth are spared the awful consequences of the otherwise inevitable conflagration. If you think that this is a "close shave," think how near it would have been should I have chosen 22 years, instead of my favored prime, 23.

In fact, this is a simplistic argument in that the effect of the impulse would not be a uniform sideways motion of 1 centimeter per second, because the asteroid is not moving in a straight line. In any case, any alteration of the orbit will lead to differing approach distances to the planets in subsequent history, as compared with the natural propagation, and thus different perturbations of its orbit. The argument does give the correct general idea of what is required, however. A fairly modest change in the velocity vector of any asteroid that is predicted to come too close to the Earth for comfort within the next century or two would be sufficient to produce a miss rather than a hit, providing that the impulse is applied early enough. I have already mentioned that we cannot predict the orbits of objects for more than about that time (a century or two), and this opens up the question of the accuracy of orbit determinations and the consequences of the uncertainties involved.

Even with orbits determined to the limiting precision of optical and radar techniques, there will always be significant uncertainty in the orbits. This is inescapable. It is similar to wanting to measure the length of a cat: If your ruler is only divided in centimeters and, in any case, the cat won't lie still, the best you can do is to determine that it is somewhere between 61 and 62 centimeters long. When one propagates an orbit forward in a computer, the deviation from reality will increase with time, because the slight uncertainties in the distances to each of the planets at any instant mean that a range of possible values for the tugs that they exert, and therefore the changes in the asteroid orbit that they induce, will be calculated. This might mean that one cannot say for sure that some asteroid is going to impact in 23 years' time, but that the best one can do is to say that it is going to pass somewhere within five times the radius of the Earth. Under such circumstances, the odds are that it will miss, but there will be a 4% chance that it will hit. In fact, if the best possible accuracy level of prediction were equivalent to five Earth radii, I would imagine that any possible passage within ten radii (at least) would be tackled. We can then relate this back to the expected impact rate: If there is one impact per 100,000 years by 1-kilometer-plus objects, then there is a one in a thousand chance of an impact within the next century; however, there is a one in ten chance of a passage within ten Earth radii in that period. This would imply that, if we make the decision to tackle any 1-kilometer-plus object predicted to pass within that miss distance, there is about a 10% chance of needing to carry out such an asteroid diversion within the next century, if we adopt a conservative approach.

Note that the previous paragraph also implies that, if you can determine only the approach distance to an accuracy of five Earth radii, you must be able to deflect objects by considerably more than that amount. That is, the precision attainable in the astronomical part of the program defines what the intercept/deflect part must be capable of accomplishing. If you are only able to deflect an object through a distance equivalent to one Earth radius, and your knowledge of the predicted distance of passage is only good to five radii, then it would be possible to apply an impulse to an object and cause it to hit, although otherwise it would have missed. Under such circumstances, there is no point in attempting a deflection, because the probability of impact would not be affected significantly; you would do just as well to keep your fingers crossed. If, however, you are able to induce a deflection equivalent to ten radii, then, knowing that the passage would have been within five radii (with a 4% chance of an impact), applying that deflection will reduce the probability of impact to zero. Obviously, to produce ten times the deflection distance of one Earth radius, the velocity change required is ten times as high; that is, our defense system would need to be able to change the speeds of asteroids and comets by 10 centimeters per second or more.

The next question that comes to mind is, "Where would you intercept the asteroid to have the greatest effect when the impulse is delivered?" The answer to that is connected with the question of the *direction* in which you might apply the impulse. In fact, rather than pushing the asteroid sideways, it is far more efficient to give it a push along the direction in which it is moving. Thus the thing to do is to accelerate (or equally well, decelerate) the asteroid by applying an impulse to its trailing (or leading) face.

As an example, we will take an asteroid with perihelion near 1 AU and aphelion at 4 AU (that is, semi-major axis 2.5 AU, eccentricity 0.6), that has been discovered, and its orbit determined to the best possible precision. It has been realized that it will pass within five Earth radii on some date unless ameliorative action is taken. Such a body varies in speed between a maximum around 37.6 km/sec at perihelion and a minimum of 9.4 km/sec at aphelion (compared with the Earth's low-eccentricity orbit, our speed altering only between 29.3 and 30.3 km/sec during the year). Changing the asteroid's velocity by 10 centimeters per second when it is near perihelion has a greater perturbative effect than doing the same thing near aphelion, with the change induced in its semi-major axis being four times as great for the former. The change thus produced in the size of its orbit does not seem very large, however, the semi-major axis being altered only from 2.5000000 AU to 2.5000531 AU. This corresponds to a distance of only about

8,000 kilometers: Given the uncertainties in any orbit determination, and the fact that this is only a little more than the terrestrial radius, is it enough?

In fact, under these stated circumstances, the asteroid would miss the Earth by much more than 8,000 kilometers. This change in semi-major axis increases the orbital period by about 66 minutes. In this time, the Earth travels about 120,000 kilometers (17 or 18 times its radius), so if the impact were expected on the next asteroid orbit (say, in four years' time), by intercepting it and giving it the 10-cm/sec shove, it would miss us by almost a third of the distance to the Moon. If the impact were due in several orbits' time, the miss distance would be correspondingly greater. The answer, then, is to intercept such dangerous objects preferably near perihelion and to either increment or decrement their orbital speeds by applying a suitable impulse—10 cm/sec being adequate (although more is better).

In passing, I will note that there is another reason for its being efficacious to intercept and divert objects when they are far from the Earth. In Chapter 5, when discussing Comet Lexell, and saw that in the latter half of the eighteenth century Lexell had two close approaches to Jupiter: The first threw it into an Earth-crossing orbit, while the second pushed it onto a long-period orbit with perihelion near the gas giant. Because close approaches to any of the planets occur rarely, this double near-passage might seem a remarkable, highly unlikely occurrence, at least *a priori*. The second close approach, however, was much more probable given the first one (that is, the *a posteriori* probability was not so small). The reason for this is that objects always return to the same position on subsequent orbits *unless* some perturbation alters that orbit. In other words, because Comet Lexell had a node near Jupiter, allowing the first deflection by that planet, it came back to that same point the next time around (when Jupiter was on the opposite side of the Sun) and again on its next orbit when Jupiter again happened to be near that point. The same thing would be true for asteroids or comets given an anthropomorphic nudge when near the Earth's orbit: They would still have a node near 1 AU and so would continue to pose a hazard on subsequent orbits. A well-planned nudge applied near perihelion, however, assuming that to be far from us, would result in the node being displaced and of no further worry, at least for a few millennia before planetary perturbations caused sufficient precession such that one of the nodes returns to 1 AU.

To deliver the necessary nudge, what sort of explosives would be appropriate? Obviously, the idea of deploying nuclear weapons in space is one that, quite rightly, would be viewed with great mistrust by the public. But would there be any alternative? The question to be addressed here is one of *specific energy*.

Many readers will be familiar with the term *specific gravity*. It denotes a type of density measurement. Rather than say that a rock has a density of 2.7 grams per cubic centimeter (or 2,700 kilograms per cubic meter), we might simply state that it has a specific gravity of 2.7. Because the density of water is one gram per cubic centimeter, anything with a specific gravity of less than 1 will float, while objects with specific gravities in excess of unity will sink.

Specific energy is a similar idea: It denotes the energy available per unit mass from a particular type of source. For example, consider a fictitious 1-gram cube of the isotope carbon-14. If we burned it with a supply of oxygen, we would get a certain amount of chemical energy released, representing a specific energy associated with atomic reactions. If we let it radioactively decay, we would get a certain amount of nuclear energy out, representing a specific energy associated with nuclear reactions; we would have a long wait, however, because carbon-14 has a half-life of 5,570 years. If it were moving at 10 kilometers per second and hit a wall, we would get an amount of kinetic energy released that represents a specific energy associated with its motion. The question here is which specific energy is highest: For a limited rocket lift capability, one wants to be able to deliver as much energy as possible to the target, which means that the specific energy needs to be maximized.

Chemical weapons (such as dynamite or TNT) can be discounted quickly, because at a sufficient speed—2.88 km/sec to be precise—*any* mass has as much kinetic energy as the chemical energy of the same mass of TNT. If we imagine a 1-ton lump of TNT being flown into an asteroid at a speed of 10 km/sec and being detonated on impact, about 10% of the energy delivery would come from the chemical energy of the TNT and 90% from its kinetic energy—to which extent, you would do almost as well to simply have had a ton of lead (or any other inert material) on board, especially since then there is far less chance of some accident occurring. Chemical weapons therefore have no role to play in this asteroid-deflection scenario; but do kinetic weapons have a use?

First of all, let us consider a small asteroid, say 100 meters in size, made of rocky material. We will "attack" that asteroid by flying a 10-ton spacecraft into it at a speed of 30 km/sec. A simple consideration[1] of the momenta involved results in the asteroid having its speed changed by around 10 centimeters per second, which our earlier considerations have indicated to be sufficient, given enough warning time. This concerns a relatively small asteroid, however. The same impulse applied to a 1-kilometer asteroid would produce a change in speed of only 0.1 millimeters per second, which is inadequate. Our options are therefore to either

1. use a much larger kinetic interceptor, with a mass of at least thousands of tons,[2] or
2. use an interceptor with a higher specific energy.

This, of course, brings us to the nuclear weaponry question. If there is just one physics equation known to any random person in the street, it is Einstein's famous mass-energy equivalence relation $E = mc^2$. To possibly repeat the familiar, E is the energy in Joules, m is the mass in kilograms, and c is the speed of light. Because c = 300 million meters per second, a lot of energy is released if one converts a small mass into pure energy. For example, if 1 kilogram were converted, the energy released would be enough to power a 100-watt electric light for almost 30 million years.[3]

Nuclear explosives, then, are intense sources of energy, allowing a vast impulse to be derived from a small mass. In terms of specific energy, a nuclear weapon's value is typically 1 to 10 million times as high as a kinetic weapon, making nuclear weaponry the only feasible choice in most imaginable scenarios. For large asteroids, even with many years' warning, nuclear explosives would be necessary; for smaller asteroids, they would again be needed if the warning time were brief (say, days to months), because large deflections would be required.

Now, how would we apply the necessary impulse? It has already been mentioned that it might be a bad move to apply too large a point explosion or impact, because that could fragment the asteroid into many large lumps, multiplying our problem. What is required is that the offending body be shoved as gently as possible onto a slightly different course, the necessary velocity change being small but the necessary energy supply being large because of the large mass involved. To change the speed of a 1-kilometer asteroid by 10 cm/sec implies a kinetic energy change of about 4×10^{15} Joules. This is equivalent to a megaton of TNT, so one might think that a comparatively modest weapon would be sufficient; however, because the coupling between the explosion and the asteroid (in terms of the nuclear energy transformed into asteroidal kinetic energy) is weak, an explosive power of many times that would be needed.

This problem was reviewed and quantified by Thomas Ahrens and Alan Harris (of the California Institute of Technology) in late 1992. They considered the full range of possible catastrophe-causing impactors (diameters of 0.1, 1, and 10 kilometers; the smallest is the minimal size to cause significant damage on the Earth's surface, the largest is about the size of the largest known Earth-crossing asteroids and periodic comets), looking into the various ways in which they could be tackled. As noted, smaller (100-meter) asteroids and comets could be diverted using kinetic techniques, but larger bodies would require alternative methods. One method might be to land some mechanism on the asteroid or comet which would then continually throw material off at greater than the escape speed (a *mass driver*), but the amount to be gradually launched from the body over many years totals thousands of tons, making this an unlikely proposition. Ahrens and Harris considered various ways of deploying nuclear charges, which might be split into three categories: (1) burying nuclear explosives in the center of the body and then blowing it apart; (2) surface-deposited charges to excavate a crater, the ejecta producing a recoil force and therefore a velocity change in the asteroid; and (3) stand-off nuclear explosions situated a few hundred meters above the body's surface, to cause a skin layer to be ejected and produce a recoil force. These will be considered in turn.

As has been mentioned previously, the idea of blowing up an asteroid or comet is not a preferred option, because many large fragments would be produced, several of which might then hit the Earth. In addition, the problem of drilling deep into the center of an asteroid presents technological problems, and the power of the nuclear weapon used is also extremely high.

The second idea, that of landing an explosive charge on the surface and excavating a crater in its detonation, is probably the most obvious. The material ejected will provide a recoil force, provided the speed of ejection is greater than the escape speed from the object (which is low for such small celestial bodies), and it is relatively straightforward to induce a velocity change of more than centimeters per second in the asteroid or comet. Ahrens and Harris were working with a bench mark required velocity change of only 1 cm/sec (whereas earlier herein I have argued that 10 cm/sec is nearer the requirement, given the uncertainties that must remain in our knowledge of the path of the bodies) and found that a surface charge of a few tens of kilotons would be sufficient for a 1-kilometer asteroid, but 100 megatons would be needed for a 10-kilometer body. The drawback of this technique is that we know little of the structure and physical strength of the objects with which we may be dealing, so a surface explosion may not only excavate a crater, but may indeed grossly fragment the object.

The preferred option of Ahrens and Harris is the third one. By exploding a nuclear weapon about 0.4 times the diameter of an object above its surface (that is, at 400 meters' altitude for a 1-kilometer asteroid; 4 kilometers for a 10-kilometer one) about 30% of its surface would be bathed in the radiation from the explosion. In particular, the high neutron flux from a nuclear weapon would be absorbed in the top 20 centimeters or so of the asteroid's surface layer. The sudden heating of this surface layer would lead to its being blown off at above the escape speed. The recoil force would be ample to induce the required velocity change in the remnant, and because the force is spread over such a wide area, there would be little chance of producing a major fragmentation. Other advantages of this method include the fact that less detailed knowledge of the object's composition, topography, center of mass, and rotation state is required. Ahrens and Harris found that the required explosive power ranges from about 100 kilotons for a 1-kilometer asteroid to 10 megatons for a 10-kilometer body, which is attainable using present technologies, although radiation-efficient weapons are desirable. The "above-surface neutron irradiation" option seems to be the appropriate one.

It was noted earlier that to change the speed of a 1-kilometer asteroid by 10 cm/sec requires a kinetic energy change of about 4×10^{15} Joules. This is about the same amount as used by a million 100-Watt light bulbs burning continuously for a year, which is a comparatively modest amount: Think how many light bulbs are kept alight by your local electric power station. The reason that nuclear explosives were deemed necessary is that a short, sharp energy delivery was envisioned. However, one could ask the question as to whether a gradual impulse delivery, lasting for a year or more, might be feasible. This idea was explored jointly by Jay Melosh (of the University of Arizona) and Ivan Nemchinov (of the Russian Academy of Sciences in Moscow). They first considered whether a giant "solar sail" could be attached to an asteroid, providing a small but continuing impulse that would divert the asteroid gradually from its Earth-impacting course. A problem with this idea is that asteroids are known to spin, so the attachment of such a sail would be virtually impossible. Melosh and Nemchinov therefore pursued a parallel idea: Why not use a vast structure like a sail as a solar flux collector to focus sunlight down onto the asteroidal surface, evaporating material which would apply a jet force as the gas expanded away from the asteroid? When they did the calculations, they found that this idea seemed feasible. For example, they deduced that such a sail/collector half a kilometer in diameter could deflect an asteroid up to 2 kilometers in size, given continuous operation for a year, and yet the reflective material would weigh only about a ton. Indeed, in most scenarios involving large asteroids or comets, they concluded that, given sufficient lead time, the solar collector concept could outperform the nuclear explosion option. Having stated all of this, such a technique has yet to leave the drawing board, and there are various objections to some of the assumptions made; nevertheless, the same could be said of the nuclear deflection concept. Obviously, we need to continue to think about this problem and have one or more acceptable solutions in hand before the time that an impactor is identified.

These deliberations were part of the thinking behind the formation, at the behest of Congress, of the Interception Committee introduced in Chapter 10. If you do discover an asteroid or comet on a collision course with Earth, how do you tackle it? This is not a new question; it has already been pointed out that numerous science fiction authors have used the idea as a main theme. In 1967, there was a now-famous exercise at the Massachusetts Institute of Technology in which the students were given the problem of diverting an asteroid (1566 Icarus, in fact), which was imagined to be on a collision course with

Earth. The conclusion of the students was that the asteroid could be diverted by using the Saturn V heavy-lift capability then available (for the Apollo program) and suitable nuclear warheads (six launches, six 100-megaton hydrogen bombs), although since then there have been criticisms of some of the assumptions made. The scenario imagined was also of the last-minute variety (an impact due the following year), requiring large velocity changes.

Unlike the Detection Committee, which met several times during 1991 and into 1992, the Interception Committee consolidated its work into a single intense meeting at Los Alamos, New Mexico, in January 1992. On the one hand, it must be said that it was a very interesting and stimulating meeting; on the other, it was also bizarre in that some of the presentations paid little regard to the laws of physics and less to any laws of economic reality. Leaving aside those talks that were clearly wildly in error (for example, in stating that the major hazard is due to objects around 4 meters in size, when we know that these explode harmlessly high in the atmosphere, and no one in recorded history has been killed by such a projectile, even though they collide with the Earth several times every year) or based on too-credulous readings of science fiction stories, the most amazing thing to me was the interface of two quite different—and yet science-based—cultures. On one hand (and in the majority at this meeting) were the military and defense scientists and engineers who were happy to propose ways of tackling the problem that ignored any budgetary constraint; on the other were the civilian (NASA, university, etc.) scientists who were accustomed to working under more stringent economic climates and therefore proposed techniques involving minimal budgets. To be fair, there were errors on each side. For example, the civilian side did not really answer the questions posed by the Congressional mandate (that is, describing a budgetary spectrum of possible responses), while the military had not done their homework on the problem (for example, not knowing that very modest velocity changes are necessary to produce the necessary deflections). To someone familiar with civilian scientific conferences, however, it was startling how little science was actually introduced. With several happy exceptions, the response of participants was not tackling the problem in hand: it was angling for new big-budget programs and the power that goes with them.

The major bone of contention between the protagonists was the relative threat posed by small, as opposed to large, asteroids: local (but perhaps state- or continent-wide) damage against a global catastrophe. The Detection Committee had already decided that the predominant hazard is due to large impactors, which cause global-scale damage. It was explained in Chapter 11 that the proposed Spaceguard program would result in the discovery and tracking of essentially all asteroids and short-period comets large enough (over 1 kilometer) to wreak such a catastrophe on the Earth. Although many smaller objects would also be found, these would represent only a small fraction of the total population of 100-meter asteroids, and one of those would be capable of causing devastation over areas as large as New Jersey, Connecticut, Holland, or Switzerland. Such objects are much fainter than their large cousins, which are the major targets of Spaceguard, and so need to come much closer by the Earth to be detected, given the design of the Spaceguard program. Consequently, one could argue (and I do) that Spaceguard should be upgraded to implement much larger telescopes—remember that telescopes using only 2- or, at most, 3-meter mirrors were proposed, whereas behemoths using 8- to 10-meter mirrors are now being built for astrophysical applications—but early on, the Detection Committee had decided to limit their recommendation to a total cost of $300 million. Generally, the members of the Detection Committee came to Los Alamos, then, with a preconception as to the nature of the problem to be tackled: large objects predicted a decade or more in advance to be Earth impactors—and let's not worry about the more frequent, smaller projectiles that might wipe out a European country.

The military-defense establishment also came to Los Alamos with a general preconception, which was totally at odds with the one just described. Their concept was that the targets were to be much smaller objects, 50 to 100 meters being typical, which would not be detected until they passed into cis-lunar space. If any one of these were on a collision course with the Earth, the warning time would be measured only in hours. This would require an in-orbit detection, interception, and deflection/destruction system: in fact, pretty much a direct extrapolation (and expansion) of the Strategic Defense Initiative (SDI) program, which was being threatened with dismemberment because of the ending of the Cold War.[4] In discussing the decision of the Detection Committee to limit its recommended program to $300 million, it was mentioned that this amount is also the approximate cost of a small space mission; thus, almost by definition, the Interception Committee, or at least a large part of its membership, was assuming a much larger budget, due to a space-based program's being envisioned.

Obviously, any interception effort would have to be space-borne, but would the sensors charged with seeking out the unknown impactors need to be space-borne as well? This question had been answered by the Detection Committee in the negative, with cheaper ground-based techniques favored, but their answer was predicated on an assumption that only bodies larger than 1 kilometer were of importance. In essence, the searching out of the impactors was beyond the responsibilities of the second (Interception) committee, which was effectively forced to deal with intercepting only those targets that the first (Detection) committee had deemed to be significant. At the initial meeting of the Detection Committee, held in June 1991 in parallel with a conference on near-Earth asteroids at San Juan Capistrano, California, the membership had been largely outraged by a paper presented by Nicholas Colella, a defense scientist from Lawrence Livermore National Laboratory. The outrage seemed to stem from an opinion that it was unacceptable to promote the development of a satellite-based asteroid detection system that would cost at least tens of millions of dollars, when the civilian scientists on that committee had been running ground-based programs for years on shoestring budgets. The mood of the civilian scientists was not improved by an after-dinner speech at the San Juan Capistrano conference by Lowell Wood, from the Lawrence Livermore National Laboratory, who was roundly booed after berating civilian scientists in general, and NASA in particular, for running expensive, inefficient, and drawn-out space missions. Essentially, he was presaging the military's getting involved in space, as exemplified the *Clementine* mission (discussed later in this chapter).

In fact, there were no military-defense scientists serving on the Detection Committee, which seems woefully short-sighted. The outrage continued when two members of the Detection Committee (John Rather of NASA HQ and Faith Vilas of NASA–Johnson Space Center in Houston) consulted defense scientists at Lawrence Livermore concerning possible detection technologies developed in the SDI program and reported back to a later committee meeting that many of those technologies looked promising, but that security requirements meant that they could not detail them. This led the Chairman (David

Morrison) to write to Rather that the latter seemed to live in some sort of "parallel universe," in that the vague information given on those possible detection techniques sounded to many as if they had been lifted from science fiction stories and did not conform to the laws of physics. On one hand, Morrison had a point: How could we assess the technologies if they were not detailed to us, and the information given seemed pie in the sky, a suspicion confirmed in some respects at the Los Alamos meeting? On the other hand, Rather had recognized that to answer the mandate handed down by Congress, it was necessary to consult a wider spectrum of scientists than those sampled from the Detection Committee membership. Even if that committee included all the available civilian scientists involved in asteroid and comet hunting, it is obvious that others who had spent the previous decade developing ways to detect (and intercept) fast-moving objects in space could have contributed some useful input.

To reiterate, the Interception Committee was, to a large extent, bound by the recommendations of the Detection Committee concerning the sort of objects it was in the business of intercepting. The various subcommittees of the former begrudgingly (in some cases) concentrated on 1-kilometer-plus objects, although some attention was still paid to potential impactors down to 100 meters in size. However, fanciful ideas of an umbrella of search-and-destroy satellites permanently stationed in orbit, ready to dash out and obliterate any asteroid coming closer than the Moon, were soundly shoved aside. Personally, I would not be at all surprised to see such ideas being reincarnated in the near future.

One argument against such an umbrella is based on economics: It would cost more than the expectancy of loss per annum. Of course, such considerations do not prevent you from spending $500 a year to insure your car, even though your expectancy of loss is only (say) $200. And the expectancy must be less than the premium, else the insurance companies would go bankrupt. Another contrary argument is that it would be exceedingly difficult to position nuclear weapons in space and still keep everyone (governments, environmental groups, peace activists, etc.) happy. There is an even stronger argument, however.

As stated previously, small asteroids, of the size that caused the devastation in Siberia in 1908, pass closer to the Earth than the lunar orbit with considerable frequency; for 50- to 60-meter objects like the

Tunguska bolide, the frequency is about once a week. Although the energy of such an object is only about 20 megatons (and hydrogen bombs of that explosive power are available), at least once a year, an object with energy in excess of the most powerful weapon passes closer to Earth than the Moon. Such an object could be used as a "superbomb" against a hostile nation, with only a relatively small impulse being needed to divert it in the desired direction. That is, any system built to render the capability of deflecting an object *away* from the Earth could also be used to divert objects *toward* our planet—specifically, toward some enemy nation.

Alan Harris and Steve Ostro (Jet Propulsion Laboratory, California Institute of Technology), Gregory Canavan (Los Alamos National Laboratory), and Carl Sagan (Cornell University) studied this problem—which they termed the *deflection dilemma*— in detail. They found that the frequency of opportunities to misuse such a system (by diverting a missing asteroid toward an enemy nation) is greater than the frequency with which you would anticipate needing to use the system to push an impactor away, depending on the abilities of that system in terms of the velocity-change it is capable of inducing. The defense system may therefore pose a larger hazard to humankind than the natural chain of events that it is supposed to protect us against; it all depends on who is in control of it.

There is therefore a strong argument, apart from the economic one, against building an interception and deflection system in advance of identifying an actual impactor hazard. That hazard could be identified by the Spaceguard survey, should it go ahead. On the other hand, there is a class of large objects that Spaceguard would likely miss or at least, of which, it would only give a few years' or months' warning: This class is the parabolic comets, as discussed in Chapter 10. I have been looking into this problem, and the contribution of parabolic comets to the impact hazard, with Brian Marsden (Harvard–Smithsonian Center for Astrophysics).

To summarize our findings thus far, Marsden and I would expect Spaceguard to most likely result in the identification, some years or decades ahead of time, of any asteroid or short-period comet due to hit the Earth, if any are indeed due within the next century or so. The construction of an interception/deflection system can therefore await such identification, although plausible techniques should be studied.

Spaceguard would not give more than a few years' advance warning of an approaching parabolic comet, however, which may be too late for the construction of a diversion system. To have no such system in place ahead of time therefore leaves us possibly unprotectable, but the alternative (building a defense system ahead of time) also leaves us open to its misuse as a hostile weapon. This is the deflection dilemma.

Obviously, merely stating that a stand-off nuclear warhead is the preferred option for asteroid or comet deflection does not, in itself, produce a viable system. There is a gamut of other considerations, such as the launch vehicle, interplanetary cruise control and navigation, homing in on the target, and so on. All of these matters were considered by the Interception Committee, and appropriate recommendations were made. At Los Alamos, probably the most useful single statement made by anyone was that of Edward Teller. He emphasized that we must have more knowledge of the characteristics of Earth-approaching objects before we can hope to tackle them in any sensible way. (Critics point out that Teller has made the same form of comment in many military-related situations, ensuring massive flows of money to the defense industry.) Take, for example, Comet Halley, a possible Earth-impactor in the distant (compared to our lifetimes) future. We have been observing it regularly for more than 2 millennia (in detail in 1835, 1910, and especially in 1986), have sent five spacecraft to study it, continue to track it using large optical telescopes (indeed, we may expect to follow it around its entire orbit), and yet we still do not know its mass to within a factor of 3. If we want to deflect it, or some similar body, we would need to know its mass because that would control the impulse needed to give it the required deviation to miss the Earth. If we know so little of Comet Halley, clearly we are even more in the dark concerning the other Earth-crossing objects that have only been recently discovered—and only studied from millions of miles away, using our telescopes here on Earth.

Consequently, it is clear that we need to investigate in detail the physical nature of the asteroids and comets that might strike the Earth, because until we understand a little more of their physical properties, we will not be certain whether a butterfly net or an elephant gun is needed to stop them. This means that we need to make *in situ* investigations of a sensible sample of those that we have detected and also to continue (and accelerate) our telescopic studies. For example, optical observations have indicated that there are several distinct compositional classes of asteroid, as shown by their differing colors. Radar data demonstrate that although some near-Earth asteroids are largely metallic (assumedly of a similar composition to nickel-iron meteorites), others seem to be rocky. We often presume that their densities are close to that of terrestrial rock (indeed, while writing this book, I made this assumption in calculating the masses, and hence impact energies, of asteroids of certain sizes), but we do not *know* this to be the case. As Teller urged, we need more knowledge, and this will be gained only by going out and taking an up-close look and likely by bringing some chunks back for investigation.

Rather than wait for some international program to come about, perhaps in a decade's time, people involved in the SDI project decided to make a start. Having developed various lightweight spacecraft technologies in that program, they very much wanted to fly a spacecraft to show that their techniques would allow them to home in on a target in space. Of course, the original aim was to be able to zero in on a satellite operated by a hostile power, but the end of the Cold War left that aim by the wayside. Back in 1985, the U.S. military hardly endeared themselves to astronomers when they used an operable astronomical research satellite, called *Solwind*,[5] as a target for a search-and-kill in-orbit exercise, destroying a useful craft when there were many inert and useless targets in space (and at the same time exacerbating the orbital-debris hazard to other satellites). One alternative was to launch their own target satellite, but that would have cost perhaps $10 million. Instead, they asked astronomers what might make an attractive natural target to home in on (but not to attack); the answer was any of the near-Earth asteroids that were accessible.

It was known that Apollo asteroid 1620 Geographos[6] was due to make a close approach by the Earth in 1994, at such a sufficiently modest relative velocity that a fly-by could be performed at a low-enough speed to make it worthwhile. Thus the Clementine project came about, with the aim of showing that interplanetary probes could be built and flown at a modest cost and with a restricted timeline. In past decades, NASA probes have typically cost hundreds of millions of dollars and have been planned and finalized years before launch (meaning that the technology used in the various instruments was often out of date even at the time of launch). The aim of the SDI team, led by Stewart Nozette, was to show that successful spacecraft could be planned, built, and launched, with the mission completed within two or three years at a cost of tens rather than hundreds of millions of dollars. On January 25, 1994, *Clementine* was launched, largely unheralded in the media, as one would expect for a military satellite.[7]

In the spring of 1994, *Clementine* completed a successful 70-day lunar mapping mission. It was intended that it then be sent on to Geographos, which would be passing close by the Earth in late August of that year. A computer malfunction during a maneuver, however, led to all of the attitude-control fuel being expended, negating any possibility of *Clementine*'s being sent on to the asteroid. At the time, we had already begun making observations of Geographos in support of the mission, so this loss was a blow. It seems that we will have to wait a little longer before we get our first up-close images of the sort of asteroids that blast holes in our planet every so often.

Although it is possible that a *Clementine II* satellite will be launched within a year or two, the only definite plan going ahead now is NASA's Near-Earth Asteroid Rendezvous (NEAR) mission. The *NEAR* spacecraft is due for launch in early 1996 and will use the gravitational assist from an Earth swing-by almost 2 years later to go on to rendezvous with the Amor-type asteroid 433 Eros in January 1999 (see Figure 14). Although Eros, one of the first-discovered Amors, has perihelion at 1.13 AU and so cannot strike our planet in the current epoch, long-term gravitational perturbations will likely bring it eventually into an Earth-crossing orbit. It is also huge, about 20 kilometers in size and thus comparable to the size of the K/T extinction impactor. Eros is therefore an excellent target for such a science mission, telling us more about the common foe we face in the sky.

If nothing else, the Clementine mission made it clear that the defense scientists in the United States are not willing to sit back and wait for something to happen, an attitude that I applaud. Nozette and his superior, Pete Worden, were instrumental in organizing an international meeting on the asteroid and comet hazard in April 1993, in Erice, Sicily. Already in January of that year, Tom Gehrels had hosted an open scientific conference at the University of Arizona, which was attended by a large number of people, with many papers presented and various authors invited to write chapters for a massive compendium that will provide a major source of information for some years to come. The

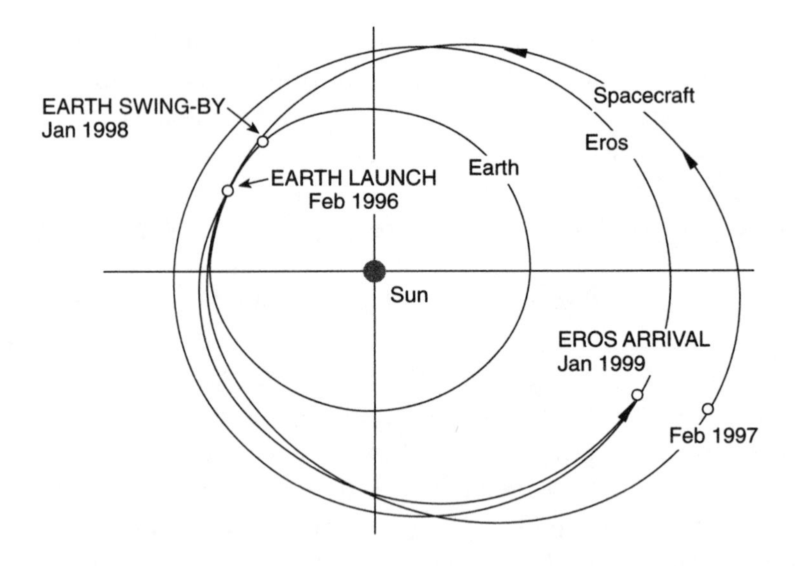

Figure 14. The flight plan for NASA's Near-Earth Asteroid Rendezvous (NEAR) space mission. A swing-by of the Earth in January 1998 will reduce the size of the spacecraft's orbit slightly, but will change its inclination by more than 10° so as to match that of the target asteroid, 433 Eros. That 20-kilometer-plus body was the first-discovered Earth-approaching asteroid (in 1898); it has perihelion about 20 million kilometers outside of the terrestrial orbit. (Courtesy of the National Aeronautics and Space Administration)

meeting in Sicily, however, was by invitation only: The aim was to get together a limited number of people involved in work on the topic and forge some sort of consensus view, so that a statement could be taken to the world's governments which would carry the weight of international opinion rather than (as is often the case) merely be seen as the lone pleading for funds by a small group of scientists.

That statement, which was argued over for a considerable time, lays out a view that already has been made clear here: that asteroids and comets pose a significant menace to humankind, that it behooves us to carry out a search program to determine whether there are any large objects due to impact the Earth in the foreseeable future, that we should study methods of diverting such a threat, that international cooperation and agreement is essential, but that it would not be wise to build or test any deflection system at the present time (due to the deflection dilemma).

13

Keeping Our Fingers Crossed

Many readers will have found the first twelve chapters of this book harrowing, given the descriptions of effects of past impacts on the Earth and their consequences, although we saw that humankind has the capability—both economically and technologically—to search out any massive asteroid or comet that may be due to hit our island home and divert it from its collision course. A global catastrophe pushing us into a new Dark Age, or worse, is therefore not an ineluctable destiny for homo sapiens. For the first time since life began on Earth, a species has the ability and the wherewithal to save itself from extinction, along with countless other species, which, the paleontological record shows us, would also perish in a gargantuan impact. Perhaps we should undertake the necessary search-and-intervention program not only for the sake of self-preservation, but also as an act of atonement for the numerous extinctions that we have inflicted on the world's flora and fauna in our exploitational activities over the past few centuries.

Such a program is possible, although we have yet to show the will to carry it through to a successful conclusion. As Greg Canavan has pointed out, the situation is similar to taking out fire insurance for your house: You do so in the hope that you will not need to make a claim. But at the moment, we not only are without insurance, we also are not able to escape from the house. We return again to the survival of our species and our cultures, in that if we had populous, self-contained colonies living in huge oases in space, or on other planets, then we could at least console ourselves with the notion that no single catastrophe could wipe humankind from the face of the universe, even if it could wipe us from the face of the Earth. To that extent, we would be guaranteed immortality. The only problem is that the establishment of such colonies may be a century away. In the meantime, we must do our best to ensure the continued inheritance of our genes, as well as achievements.

This is the lesson that we have learned in recent years, as we have gained an understanding of the larger impacts on the Earth and how they affect the environment. As is expostulated in Chapter 7, however, it is feasible that the hazard posed by smaller asteroids and comets less than a few hundred meters in size has been grossly underestimated by the majority of research workers; indeed, I argue that the very nature of that hazard has been misunderstood. Under those circumstances, the suggested Spaceguard survey would not provide any measure of protection or certainty against the predominant risk. Having stated this, it is clear that Spaceguard would allow the true nature of the hazard to be ascertained. Already, the data from the Spacewatch telescope has provided evidence that the view held by myself and a few collaborators—that giant comets break up so as to produce coherent streams of objects ranging from meteoroidal to asteroidal dimensions—is correct, and there is at least one such complex currently circuiting in the inner solar system. I have described the complex that we have identified as being associated with Comet Encke; however, there is also a similar grouping seemingly linked with the large Apollo asteroid 2212 Hephaistos and Comet Helfenzrieder (which, mysteriously, was observed only in 1766, not being seen again, despite its having an orbital period of only a few years). In addition, the evidence is growing that the Kreutz group of sungrazing comets is derived from a giant comet's disintegration, producing many smaller bodies that can strike the Earth at very high speeds.

If this view is correct—that impacts by smaller objects like the

Tunguska projectile generally occur in clusters rather than randomly in time—then the idea of protecting humankind from impacts must take on an altogether different guise. For such small bodies, it would not be possible to give warnings years in advance, the time frame being closer to hours. A constant patrol of cis-lunar space would be needed, with multiple intercept vehicles maintained in readiness, most likely in geocentric orbit. This would be much more expensive than the scenario sketched out in Chapter 12, but it would be necessary should we decide that we cannot afford *not* to tackle the problem. In fact, it is similar to the approach considered by the Star Wars personnel when they first addressed the topic: batteries of nuclear-armed satellites ready to attack any intruding asteroid at a moment's notice. We have also seen, however, that such a readiness would actualize a new danger, due to the possibility that such a defense system could be used offensively, this posing the deflection dilemma.

All of these matters may seem too futuristic for consideration in the twentieth century, but we have reached the stage where they need to be discussed openly and not just academically. In Arthur C. Clarke's recent novel, *The Hammer of God*, he gives an exemplar of how the people of the future might be polled in order to decide whether to try to deflect an asteroid found to be on a path that might hit the Earth, although surety was not possible in that fictional case, as indeed would likely be any real-life scenario. In educating the general public with regard to the possibilities of the coming decades, futuristic writers have performed a great service by sensitizing people so that the progress of science and technology are more easily accepted. Although the Spaceguard plan takes its name from Clarke's earlier fictional program in the novel *Rendezvous with Rama* published in 1973, such far-reaching suggestions were in the literature even two decades earlier. For example, consider the following excerpt from *Target Earth*, by Allan O. Kelly and Frank Dachille, which was published in 1953 (and note that the concept of the communications satellite was invented by Clarke a few years before that):

> Therefore, it behooves us to consider ways and means to ward off a Day of Reckoning that may be set up in the mechanics of the Solar System and the Universe. In the increasingly numerous scientific discussions of rockets for interplanetary travel, and in the consideration of man-made

> satellites or artificial moons, we see the beginnings of a system for the protection of the Earth. This system will require perpetual surveillance of a critical envelope of space with the charting of all objects that come close to a collision course with the Earth. It will require, further, that on the discovery of a dangerous object, moves be made to protect the Earth. To this end might be used rocket "tug boats" sent out to deflect and guide the object from the collision course.
>
> This patrol may be achieved by the use of mechanical (or man-controlled) satellites, ranging well out beyond the orbit of the moon. The everlasting night of space would permit sensitive scanning of the critical envelope, while the distance from the Earth would allow a valuable time factor. Guided rockets of proper design, dispatched from the most convenient satellites, would then serve to intercept, and to deflect, dangerous objects. All this calls for the ultimate scientific skill of man, but not more so than that skill required for the development of rockets for interplanetary exploration, and certainly not as complex as that required for the interstellar exploration and colonization discussed in a serious journal.
>
> We see then that the personal safety of civilized man extends outward from the police powers in his home town to a full and vigilant patrol of outer space.

Despite four decades of advance in our knowledge, one would be hard-pressed to write a better brief desideratum regarding the protection of humankind from the asteroid and comet hazard than this depiction from *Target Earth*. Many of the details have changed—for example, we know that it is much cheaper to survey space from the Earth, though perhaps expense is no object—but the sentiment expressed by Kelly and Dachille has been consolidated rather than weakened in the intervening period.

Having brought up the subject of expense, we could perhaps close with a comment about money itself. Unlike the currency of the United States, for which the term *greenback* is proverbial, Australian banknotes exhibit a veritable rainbow. Each has its own color, often garish, and each is of a different size, the dimensions ascending with the face value. I should also note that Australia seems to be the first nation to introduce durable plastic currency in the form of its $5, $10, and $20 bills, perhaps reflecting the fact that the inhabitants like to spend time on the beach; I have often appreciated being able to lock my wallet in the car and go for a run along the shore, then take a cooling dip in the ocean, fully confident that the $5 in my pocket will still be presentable to obtain refreshment.

Australian banknotes also reflect another aspect of the antipodean culture, however. A country with a population of only 17 million people cannot be preeminent in all fields of science, but if one were to have to choose a handful of areas in which Australia's achievements have been exceptional, then medical research, biochemistry, agricultural science, and astronomy form a strong suite. This is demonstrated by the images used to illustrate the banknotes, those with astronomical connections being of interest here. The $100 bill has on one side a picture of John Tebbutt, a nineteenth-century comet discoverer who had an observatory at Windsor, just northwest of Sydney. The $50 bill has an impression of the massive Parkes radio telescope—often used to receive data from NASA spacecraft and other satellites—along with plots of interstellar regions mapped with that instrument, and a trace of the pulses of radio waves from a pulsar. Each banknote has an official theme, stating what had inspired its design; the theme for that $50 bill begins:

> Endless research leading to the survival of man on Earth through greater knowledge of outer space . . .

When the note was first issued two decades ago, the folks at the Australian Mint could not have known just how true that sentiment might be. One hopes that one day soon the Australian government, along with the governments of other nations, will realize that survival on our island in space does indeed require a greater knowledge of outer space, but of objects much closer than the nearest pulsar. It is not a matter of *if* an asteroid or comet will hit the Earth with cataclysmic consequences, it is merely a matter of *when.*

Epilogue

The Crash of '94

Notwithstanding the Tunguska event of 1908, which was a damp squib compared to the ferocious assaults that asteroids and comets can unleash, humankind has not had a chance to view a massive impact on the Earth during recent times, or I would not be here to write this book. You will all be aware, however, that another planet *has* suffered such a collision. This Epilogue, written in November 1994, post-dates the series of cataclysmic explosions on Jupiter which we all witnessed—on TV if not at a telescope—last July. Those distant explosions are proving to have repercussions for us here on the Earth, in that they proved that scientists were not joking when they talked about the power of comet and asteroid impacts—my warnings earlier in this book are fact, not fiction.

In this concluding chapter there are three matters to be addressed. On the side of science, I will discuss the lessons that we have learned from this opportunity—and in particular its advance prediction—with regard to impacts on the Earth, and thus safeguarding humankind from similar cataclysms. Second, there is a political angle, in that there has been a response from various governmental organizations—the United States Congress, the Council of Europe, some action in Russia, and a briefing document produced for Australian parliamentarians, to list a few. Third, another aspect, which could also be considered to be politics, has been the response of astronomers to media coverage of not only the impacts on Jupiter, but also, more importantly, the possibility of impacts on the Earth.

But first to science. The jovian impactor making the news was called Periodic Comet Shoemaker-Levy 9. Now, that name is one huge mouthful, and I could refer to it with the astronomers' shorthand title of *1993e*—that is, the fifth comet discovered in 1993—but most readers will more readily recognize it as *S–L9*. This comet was discovered by Gene and Carolyn Shoemaker and David Levy in March 1993, hence the two surnames attached to it. The inclusion of the number 9 derives from the fact that, previously, this team had discovered eight other periodic comets. In addition, over the years they have jointly found more than twenty parabolic comets (each of which is called Comet Shoemaker-Levy, except with distinguishing notations, such as 1994d), plus several other individual discoveries, including Comet Shoemaker 1994k. Periodic Comet Shoemaker-Levy 9 is the phenomenon, then, that we will call S–L9.

When it was discovered, instead of being a single body it was a chain of individual cometary nuclei stretched out along a line millions of kilometers in extent. Soon after it was found, a backward-extrapolation of its motion indicated that these nuclei had made a very close passage by Jupiter in July 1992, and it is clear that the tidal stresses induced at that time on a single larger body caused a fragmentation and consequently the chain discovered eight months later. Why was it not found earlier? The answer is that, generally, astronomers avoid taking photographs near Jupiter in searches for asteroids and comets because the exceptional brightness of that planet fogs the emulsion over the entire frame: These are few celestial objects brighter than Jupiter. S–L9 was found when it had moved far enough away from Jupiter to appear in an image uncontaminated by sunlight reflected by the planet's bright cloud decks. Because Schmidt telescope images cover several degrees across the sky, S–L9 was not found until it was that distance away from Jupiter. When it was spotted, it was near to its maximum separation from the planet; if it had a slightly smaller jovicentric orbit, it might never have been discovered at all. Perhaps we have a lesson to learn from this. In fact, a team of Swedish and Uruguayan astronomers had suggested just the year before that the region close to Jupiter might be a fruitful area to scan for comets. They were soon proven correct, although not in the way that they had imagined.

Looking forward in its orbit, calculations made right after S–L9's discovery indicated the likelihood that these nuclei would impact Jupiter in July 1994, producing a jovian fireworks display. Of course, this caused a buzz immediately, solar system astronomers eagerly anticipating what might be seen as a result. It is believed that the well-known rings of Saturn were produced when another comet—or perhaps several at different times over the past 100 million years—passed too close by that planet and was broken up, with some of the material being accumulated in saturnocentric orbit while the rest remained on path circuiting the Sun. Jupiter, Uranus, and Neptune also have tenuous rings, assumedly produced in the same way. The physics involved in the breakup of such bodies is very simple, and it is also the reason that planets do not have any large natural satellites or moons within two or three times their radii: Any large, weak body passing that close is torn apart by the differential gravitational attraction between the face closest to the planet and that furthest away. The question was asked, then, whether Jupiter would acquire an extensive ring system. Many other plausible ideas came to mind: Would the impacts produce a new Great Red Spot, a series of red spots, or maybe white spots like those scattered across the planet?

Obviously, there were many predictions to be made—some wild and some sensible. Those which turned out to be correct—there were few of them—will be remembered, while the majority that were wrong are already almost forgotten. For the record, ahead of time I told people, through the mass media, that I expected the impacts to be a fizzer, but I hoped that I was wrong. And wrong I was.

My main reason for discussing S–L9 is that the timetable of events surrounding its discovery and the prediction of impacts on Jupiter provide an exemplary lesson in what might happen should we find a body on a collision course with the Earth. There are certainly many differences—for example, S–L9 was not in a heliocentric orbit, and also Jupiter poses a much larger target, especially when one takes into account gravitational focusing—but still the events occurring after the comet chain was discovered provide an interesting illustration of the steps taken by astronomers in identifying a future impact.

The story begins on March 24, 1993. The Shoemakers plus Levy were obtaining their monthly set of photographs of the heavens, using the small Schmidt camera at Mount Palomar in California. Whenever they (or anyone else) finds an important celestial object—such as a comet, supernova, or other suddenly appearing phenomenon—it is announced to the astronomical community through the single-page *International Astronomical Union Circulars* (IAUCs), prepared by a small team under the direction of Brian Marsden at the Central Bureau for Astronomical Telegrams (an anachronistic name), which is located at the Smithsonian Astrophysical Observatory in Cambridge, Massachusetts. The IAUCs are still sent out by mail to some subscribers, but nowadays most observatories and universities receive them immediately by electronic mail. On March 26, Marsden issued IAUC 5725 in which he stated, "Cometary images have been found [by the Shoemaker-Levy team] . . . the appearance is most unusual in that the comet appears as a dense, linear bar . . ." He went on to say that the object had been confirmed by Jim Scotti using the Spacewatch camera in Arizona, with at least five distinct condensations being identified. Marsden concluded by writing, "The comet is located some 4 degrees from Jupiter, and the motion suggests that it may be near Jupiter's distance." A peculiar object, a new member of the celestial zoo, had been found, and although little was known of its nature at that time, it did seem that it was not far from Jupiter—suspicious.

By the following day, further observations had been obtained from teams at the Oak Ridge Observatory in Massachusetts and at the University of Victoria in British Columbia. In IAUC 5726, Marsden was able to determine an orbit that would imply that the comet had an approach to Jupiter (at most 6 million kilometers distant) in July 1992, which is close on the astronomical scale but not near enough to cause a tidal disruption. But then, the separation may have been much less, more astrometric data being needed in order to tell. Already in the minds of astronomers, it was clear what had actually happened; Marsden wrote that ". . . the comet would have been as bright as its Feb. 1992 opposition as it is now [and so would have been discovered then], but splitting presumably occurred near Jupiter (as with P/Brooks 2 in 1886), and this is presumably to be associated with brightening." Photographs taken of the vicinity of Jupiter in early 1992 by a team from the University of Uppsala in Sweden were inspected and found to show no trace of the comet, indicating that it was dimmer the previous year by a factor of at least 100,000. The comet may have been there for decades, but dark and inactive, now appearing an obvious object because it had split and exposed a large amount of dust to sunlight.

The hook was now well baited, and many other telescopes were turned toward this remarkable new phenomenon. In Hawaii, David Jewitt and Jane Luu interrupted their search for trans-Neptunian objects to look at S–L9 and found that there were at least 17 distinct nuclei in the train. They coined the now-famous description that the nuclei were "strung out like pearls on a string." If there were indeed many separate nuclei, then one could argue that these should be considered as separate comets, and so the Shoemakers and Levy should be credited with many discoveries; but because the decision had already been made that this constituted a single discovery, Carolyn Shoemaker will have to wait a little while yet—although probably less then a year or two—before she overtakes Jean Louis Pons (an early nineteenth-century French comet aficionado) as the most prolific discoverer of comets in the history of astronomy.

While more observations came flooding in, prompted by the excitement, other astronomers went back to photographs they had taken in the vicinity of Jupiter in the previous weeks to look for images that had been overlooked at that time. Two Japanese amateur astronomers, Messrs. Endate and Otomo, found the comet on their photographs taken on March 15 and 17. A Venezuelan observer by the name of Naranjo made an independent discovery on a photograph he took on March 24, a few hours before the Shoemaker-Levy team obtained their discovery shot. The team under Eleanor Helin, which also uses the small Palomar Schmidt for asteroid and comet searching, detected the comet train on March 19 and 21, noting the images but not following up on them to confirm their cometary nature. From personal experience, I know that searching films for comets is exasperating at times, because asteroids give distinct, crisp trails whereas comets produce fuzzy images that are replicated on almost every film by flaws in the emulsion. Thus a bizarre image, a bar with a blurred appearance, would quite likely be rejected as being yet another flaw unworthy of further attention. The moral of the story is that the laurel (and thus the comet's name) goes not to those who detect the object first, but to those who recognize and communicate their discovery to the appropriate authority before any other claimant.

By April 3, Marsden was able, using the wealth of positional measurements then in hand, to show that S–L9 was almost certainly in an orbit at least temporarily centered on Jupiter; that is, it seemed that the comet must have originally been in a heliocentric orbit, but a chance approach to Jupiter at just the right speed led to the planet's refusing to release it from its grip. Even now, the history of the orbital evolution of S–L9 remains a mystery: This capture may have been in just the last few years, but it could have occurred a century or more ago. Marsden noted that a tidal break-up in 1992 would have required passage within 150,000 kilometers of Jupiter, which is little more than twice the radius of the planet. Within a short time, there were sufficient accurate astrometric observations to allow a precise backward integration of the orbit—a difficult task because both Jupiter and the Sun imposed strong pulls upon S–L9—this showing that the progenitor did indeed pass perijove 110,000 kilometers from the center of the planet (or 40,000 kilometers above the cloud tops) in July 1992.

The main game was shifting, though, from what had happened in the past to what was to happen in the future. By the middle of May, various researchers had come to the conclusion that at least part of the train of fragments would strike Jupiter in July 1994. On May 22, Marsden had issued IAUC 5800 in which he noted that orbital solutions were indicating a minimum distance from the center of Jupiter that was less than the radius of the planet, which is a dry way of saying that an impact might occur. On the same day, he wrote in IAUC 5801 that an impact seemed likely but that "it must be emphasized that a 1994 collision of the train center with Jupiter is not assured"—a suitably conservative line. Apart from Marsden, Andrea Carusi of the Italian National Research Organization, and also Don Yeomans and Paul Chodas of the Jet Propulsion Laboratory, found it likely that the center of the train, at least, would impact Jupiter; a few stragglers might escape, but most of the nuclei would probably hit the planet.

These suggestions were still based only on probabilistic arguments, not definitive knowledge, because the observational uncertainties allowed the possibility that the train would miss. But during 1993, more and more precise positions were measured, and with the increasing time span and the decreasing wait until the next perijove, the degree of uncertainty dropped until by November 22 Marsden was able to state in IAUC 5893 that "it is absolutely certain that all of the individual nuclei" would hit Jupiter in July 1994. Scientists are a conservative bunch, and if such an experienced astronomer as Marsden says an impact is certain, you can bet your lifesavings on it. There was no longer any doubt. The rest, as in the hackneyed phrase, is history.

Few can have missed seeing the spectacular images of the impacts obtained from observatories around the world, and from the Hubble Space Telescope, showing that the results of the impacts were far from being a fizzle. The full analysis of those images, and other types of data, will continue for some years: If astronomers had no idea what to expect ahead of time, it would be unreasonable to demand any instant answers. Indeed, more than three months later, data collection continues with various radio telescopes—Jupiter is still emitting at an enhanced level—and the dark spots produced by the impacts may be smeared out, but they have certainly not yet disappeared. Data are also still being returned form the *Galileo* spacecraft, which obtained direct images of the impact region; that region could not be viewed directly from the Earth.

One important question that remains a puzzle is the nature of the impactor. We have been referring to it as having been a comet, but it displayed none of the definitive characteristics of a comet. Certainly there was dust around the train of nuclei, but, at such a great distance from the Sun where the solar energy is too weak to cause much volatile evaporation, there was no sign of a normal gaseous coma. The dust could therefore be merely the detritus from an asteroid breaking up near Jupiter. It was expected that in the impacts, the signature of water would be detected, if the object were a comet, but again the results have been negative. On balance it seems that this object should really have been classed as being an asteroid. My own view is that this underlines a point that I made earlier in this book: Asteroids and comets are basically the same phenomenon, exhibiting a similar range of characteristics as do the animals we call mammals, that class encompassing mice, kangaroos, humans, elephants, and whales.

During the media furor at the time of impacts, many times the fireworks on Jupiter were described as being the chance of a lifetime. It was a surprise to many of us that relatively small telescopes—certainly any with an aperture of more than 10 centimeters—could be used to see the dark spots on the face of Jupiter, defying those scientists who had predicted new *bright* spots. I went away to a few reference books and found that similar dark spots have been reported several times in the past century, in 1885, 1928, 1939, and 1948, for example, meaning that perhaps comet impacts on Jupiter are not really a once-in-a-lifetime experience, but occur more frequently than we had thought. It is just that the impacts of July 1994 had been predicted, and intensely observed.

It did not take governments long to recognize that we were serious when we spoke of the degree of catastrophe that would be wreaked on the Earth should such an impact occur here. Scars on Jupiter four to five times the diameter of our own planet were testament to that. The first impact occurred on July 16; on July 20, the U.S. House of Representatives wrote the following into the NASA Authorization Bill:

> To the extent practicable, the National Aeronautics and Space Administration, in coordination with the Department of Defense and the space agencies of other countries, shall identify and catalogue within 10 years the orbital characteristics of all comets and asteroids that are greater than 1 kilometer in diameter and are in an orbit around the sun that crosses the orbit of the Earth.

The text continued by requiring the NASA Administrator to submit to the Congress a program plan, including budgetary requirements, by February 1, 1995. As I write, the committee formed to accomplish that task, chaired by Gene Shoemaker, is preparing that plan. As to what that committee's response will be to the stipulation of 10 years as the time frame—we have earlier seen that at least 20 years would be needed, and the laws of economics cannot overrule the laws of celestial mechanics—we shall have to wait and see.

It was not only in the United States, however, that political action stirred. In Europe, a motion (which is still under discussion) was put to the Parliamentary Assembly of the Council of Europe in Strasbourg. That motion encouraged the signatory nations—31 countries are members of the Council of Europe—to join with the United States in formulating a plan of action to tackle the million-megaton menace. In Russia, a conference was held in September 1994 which developed a plan for what the participants term the *Space Protection of the Earth.* In Australia, a governmental briefing document was produced within weeks of the impacts on Jupiter, that document being specifically designed to advise politicians on the appropriate steps to take regarding the hazard of impacts on the Earth. Judiciously, it advises that a telescopic search such as the Spaceguard Survey should be carried out, with Australia playing a prominent role, but that at the present time it is not appropriate to begin building space hardware for asteroid or comet interception and diversion. Such a step should await the identification of a specific hazardous body.

Having witnessed the phenomenal crash of a comet into Jupiter, what are the lessons that we might learn vis-a-vis possible future Earth impacts? The process that astronomers went through in determining that an impact was likely, achieving certainty as sufficient precise data were acquired, provides a salutary lesson for Earth impacts. People often ask me questions along the lines of "What would you do if you found a comet on a collision course with the Earth?" In posing that query, they imagine that we peek through a telescope and spot one coming our way, then drop everything to head for the hills (as if that is of any use). That would not be the process at all: It would take at least months of careful observations to determine a comet's or asteroid's path with sufficient accuracy to state with any degree of confidence that an impact was likely. If the impact were due soon after discovery—within months—then it would be too late anyway. Another question often asked is, "Would the news get out? Would the government allow people to know?" Let me assure you all that even if one government tried to muzzle its astronomers, with the worldwide rapid communications that we enjoy, indeed that are essential for our work, the news would soon be spread. The telephone and fax numbers for many news organizations sit next to my office phone and computer terminal.

And that brings me to my next point, concerning interaction with the mass media. Let us return to my description of the concepts of coherent catastrophism as opposed to stochastic catastrophism (impacts occurring randomly in time). The impacts by the fragments of Periodic Comet Shoemaker-Levy 9 on Jupiter are an extreme example of the coherent catastrophism idea. For the Earth, my coworkers and I visualize large comets in short-period orbits with perihelia less than 1 AU, and aphelia near Jupiter—in fact, an orbit like most of the originally long-period comets recently trapped by Jupiter into small orbits. Such comets make frequent close approaches to Jupiter, and every so often they will stray too close, becoming shattered and spreading their fragments into a stream with basically similar orbits. Equally well, thermal stresses (or other factors that we do not yet understand) can cause break-ups in space, far from any planet. Certainly S–L9 broke up, except in a jovicentric orbit, and we witnessed several of its fragments becoming too dim for detection (that is, they became asteroidal) even within the two years after the disintegration. The event was a microcosm, showing what could happen on a grander scale in the inner solar system. When a giant comet fragments on a heliocentric orbit, as one must from time to time, the flux of impactors on planet-crossing orbits is enhanced many-fold, marking an epoch of great danger for life on Earth. Because the fragments soon became dispersed from their original compact train (as we saw occurring for S–L9 even within two years), the impacts on the Earth would continue not for just a week, but spasmodically over the next 100,000 years.

The opponents of this concept say that such an event is unlikely. If one wanted to argue on probabilistic grounds, one could revert to the case of Periodic Comet Shoemaker-Levy 9. Prior to 1993, the calculations—performed by me among others—implied that cometary impacts on Jupiter occur about once per century. Thus if one had asked, in early 1993, the question, "What is the probability that there will be 21 cometary impacts on Jupiter within one week next year?", the answer, based on our state of knowledge as it stood, then, would have been, "one part in 10^{70}, quite likely rather less," and that is too low a chance to take seriously. By the end of the year, the answer to the same question was, "The probability is identical with unity: The impacts are certain."

Simple probabilistic arguments cannot be used in our present state of ignorance, there being so many things that we simply do not understand. One might quote the famous aphorism that the universe is not only stranger than we imagine, it is also stranger than we *can* imagine. In early 1993, multiple impacts on Jupiter were not imagined as something to look out for, and the crater chains on the jovian moon Callisto, found using the *Voyager* spacecraft, were a real puzzle. In late 1993, such impacts were not only being eagerly anticipated, but also Callisto's crater chains had been interpreted as being a legacy of many similar comet break-ups in the past.

So to the mass media. During the impacts on Jupiter, the question was often asked, "What can we learn from this?" To some extent I have already answered that question. Another very significant point, however, is that broken-up comets are especially dangerous. The conventional view is simplistic, seeing monolithic bodies coming from outside the planetary region of the solar system, or from the asteroid belt, and by mischance running into the Earth from time to time. My contrary opinion is that shattered comets pose a special risk that dominates the hazard. That opinion is scorned by most of my colleagues, who say that comet fragmentations are too infrequent to be of concern, to which extent what happened soon after the explosions on Jupiter will be of interest to the reader.

Three weeks after the final jovian impact a renowned Californian comet searcher, Donald Machholz, found another comet (now known as P/Machholz 2, or 1994o) to add to his creditable list of discoveries. (As an aside, the reader might like to note that this comet was not in a region of the sky that would be scanned by the Spaceguard Survey.) A few weeks later, in late August, another amateur astronomer found that this comet was not alone: It had a companion nucleus some distance away, indicating a fragmentation event had occurred some time ago, perhaps as much as a decade or two. This interested me because P/Machholz 2 had by then been determined to be on an Earth-crossing orbit with a short period, coming back every six or seven years. By Tuesday, September 6, another three nuclei had been found, making a total of five, and since then, one more has been added. As I write, we are planning observations using a larger telescope than has been used to date, to see how many fragments actually exist.

To me, this was an obviously newsworthy event. The last short-period comet to fragment on an Earth-crossing orbit was P/Biela in the middle of the last century. How that comet produced phenomenal meteor storms in 1872 and 1885 has been mentioned earlier in this book. My computations showed that P/Machholz 2 has an orbit that currently does not come closer than about 17 million kilometers to the Earth, so that an impact in the foreseeable future is not possible. At that early stage, the orbits of the fragments were not very well known, but I could be sure that nothing dramatic would occur within a century or so. The problem is that either news is hot or it's not; one either makes a story public, or else keeps quiet and is then condemned for not letting people know about an interesting and topical event. On September 7, I did a radio interview with the Australian Broadcasting Corporation about this new fragmented comet, and the next day the story was carried in *The Australian*, the national newspaper.

I often supply information to Robert Matthews, a science journalist with *The Sunday Telegraph* in London. Matthews is a particularly good contact because he conducts his own astronomical research (see note 3 of Chapter 6), and is therefore conversant with the intricacies involved. After I spoke to him, Matthews wrote a story that was carried in his newspaper on September 11. He correctly quoted me as saying that neither the comet nor its fragments could hit the Earth in the next century according to calculations made to date, but that continued surveillance was required in case we were wrong. That story was then picked up by Reuters, a shortened—and, of course, sensationalized—version being printed in newspapers around the globe. This outraged many solar system astronomers who were caught on the hop, not having kept up with events of the previous two weeks. Suddenly, radio and TV stations were calling them, asking about this new fragmented comet. I found their naiveté in dealing with the press to be staggering, because it is always to be expected that the media will stretch stories as far as they can in order to grab their audience.

But there was another reason for the outrage of these astronomers, and that is that the idea of shattered comets does not fit in with the concepts they have advocated for protecting humankind from impacts. The Spaceguard concept is based on an assumption that single, monolithic bodies pose the major hazard to our species and other life on planet Earth. A rather different form of hazard is that resulting from a stream of cometary debris that the Earth passes through, delivering perhaps dozens of Tunguska-type explosions. The impacts on Jupiter were, in some ways, an embarrassment to those who would have you believe that single large asteroids and comets are the things about which to worry. To immediately find a shattered comet (P/Machholz 2) that might eventually deliver a series of blows to the Earth could lead people to doubt that we really understand what is going on. Fragmentations of comets are frequent events, and we will observe more now that astronomers are sensitized to the phenomenon. On October 4, Jim Scotti, using the Spacewatch telescope, found that Periodic Comet Harrington had broken into at least three parts. This comet has perihelion outside of the Earth's orbit, making terrestrial impacts impossible, or I would have been on the phone to the media again.

Comparing the hazard posed by intact comets and asteroids with the numerous fragments produced when they break apart is beyond us in our present state of ignorance. If you were locked in a dark room with a homicidal maniac with a firearm, would you rather that he had a rifle or a shotgun? The latter is more likely to hit you, but perhaps with less catastrophic consequences. The Spaceguard Survey would most likely lead to our being able to solve the problem of what the dominant hazard might be and to decide how to protect ourselves against future impacts. In the meantime, we need to keep hoping that the trigger is not pulled.

Notes

Chapter 1

1. TNT, or trinitrotoluene, is a form of high explosive that is usually used as bench mark for the energy released in a particular detonation. For example, the nuclear bomb dropped on Hiroshima had an explosive power equivalent to about 13,000 tons of TNT (13 kilotons). One ton of TNT is equivalent to about 4.15 billion Joules of energy; the Joule is the metric unit of energy, a power of one Watt implying an energy supply of one Joule per second. For shorthand purposes, energy equivalent to one megaton (one million tons) of TNT will be abbreviated as Mt throughout this book.

Chapter 2

1. Simon Laplace should, however, be credited with being the person who first advanced the idea of a Black Hole, although his concept predates the discovery by Einstein of the gravitational theory which is properly needed to formulate the solution to the Black Hole phenomenon. Laplace (1749–1827), whose full name was Pierre-Simon, Marquis de Laplace, also made many other seminal discoveries in mathematics and astronomy.
2. It is worthwhile to point out here that similar reasoning applies to impacts by asteroids or comets on the Earth. I am often asked whether such an impact would knock the Earth out of its orbit. The mass of a typical comet is less than one part in a hundred billion of the mass of the Earth, so it impedes the Earth in its progress about the Sun by about the same amount as a gnat slows your car as it splats into your windshield.
3. In calculating the probability of any particular comet or asteroid hitting the Earth we begin with the fact that our planet's radius is just less than 6,400 kilometers, and our orbit about the Sun is a near-circle with a radius of 150 million kilometers. Thus the cross-sectional area of the Earth—the target area for an impact—is one part in $[4 \times (150{,}000{,}000/6{,}400)^2]$, or one part in 2.2 billion, of the area of a sphere of radius 1 AU (the surface area of a sphere is

$4\pi r^2$, the area of a circle is πr^2). Because the comet gets two chances to hit the planet—once on the way in toward the sun and once on the way out—a crude estimate of a general collision probability is one in 1.1 billion, which we will call one in a billion as a round figure. Because most objects of the types in question have orbits confined to areas near the plane of the planetary orbits, they cannot cross far from that plane, which reduces the area in question by a factor of 3 or 4. Therefore, a reasonable estimate for the typical collision probability with the Earth is one in 300 million. In fact, a proper calculation of the collision probability is much more complicated than this (for example, the Earth's gravitational field can drag in objects that would otherwise miss), with widely disparate values resulting, but this simplistic calculation gives a good rule of thumb that is in broad agreement with the values derived using complex algorithms.

4. Note that the tail of a comet does not lag behind its motion; comet tails are directed away from the Sun.
5. Icarus was the son of Daedalus in Greek mythology. They fled from their jail in a tower by constructing wings with feathers held together with wax. Despite his father's warnings, Icarus flew too high and near to the Sun as they made their escape, with the result that the wax melted and Icarus plunged to his death. Astronomers like to pick appropriate names for their discoveries; there is also an asteroid called 1864 Daedalus.
6. The prefix *P/* is astronomical shorthand for "Periodic comet."

Chapter 3

1. Why else would *tsunami* be one of the most common Japanese words adopted into the English language, along with (due to recent fads) *sushi*, *kimono* and *futon*? Literally, *tsunami* means "harbor wave." The commonly-used alternative—*tidal wave*—is incorrect because tsunamis are not caused by the tides (that is, variations in sea level due to the attraction upon the oceans of the Moon and the Sun).
2. A disproportionate number of meteorites are nickel-iron conglomerates because they are more likely to survive entry. Some near-Earth asteroids that have been detected by radar by Steve Ostro (Jet Propulsion Laboratory, California) have reflectivities indicative of their being metallic, and the optical colors of many asteroids also suggest that they are metallic in nature. Examples of craters formed by metallic impactors include Meteor Crater in Arizona and the Henbury complex of craters in central Australia; in fact, most terrestrial impact craters smaller than 2 kilometers in diameter seem to have been created by metallic bolides.
3. This is an underestimate, because we are considering only the deaths from impacts at the threshold for global effects. We should really add in the 1,000

or so deaths from 100-meter impactors, and so on, up through the size spectrum to determine a net value from all impacts large and small.

Chapter 4

1. Another example of such a phenomenon is the explosions that occurred from time to time in nineteenth-century flour and cotton mills due to the amount of dust in the enclosed, poorly ventilated spaces. Again, charge build-up and consequent discharge—lightning—was responsible.
2. In fact, samples that can be definitely tied to originating from the Chicxulub site, and dated as having been shocked at the time that the crater was formed, have been found in various places in the United States.
3. I used the term *so-called* because this is not really the way in which a greenhouse works, the warming of a greenhouse being mainly due to the trapping of the air within rather than the action of glass in restricting the transmission of different wavelengths of light (that is, the greenhouse warms due to the suppression of convection, rather than radiation). By the way, contrary to popular opinion, the greenhouse effect is not a recent idea; its first correct scientific explanation was given by Charles Babbage in 1847.
4. Note that the unit is the Kelvin; there is no such things as a "degree Kelvin." Use of such a fictitious unit is a mistake often made, by astronomers in particular.
5. "Ah, hah!" says the attentive reader. That wavelength is in the green part of the visible spectrum, so the Sun should appear green. The fallacy here has to do with our eyes—essentially, they are energy detectors, not wavelength detectors. The energy of a photon is given by its frequency multiplied by a certain number (Planck's Constant). Although one might think that because one multiplies the wavelength by the frequency to get the speed of light, everything settles out. But it's not so simple, I'm afraid. The varying numbers of photons emitted by some hot object are governed by the Planck Distribution. It is a *distribution*, not a *function*. This means that one can say only how many photons (or how much energy) is emitted per frequency interval (or per wavelength interval). Finding the peak of the distribution in the wavelength domain, one derives the equation given, but in the frequency domain, the constant one determines is quite different. It is not as simple as substituting for the speed of light. The bottom line—your eyes are not deceiving you, the Sun is indeed yellow.
6. The solar flux at 1 AU—the *Solar Constant*—is about 1.35 kilowatts per square meter at the top of the atmosphere.
7. Obviously, *noxious* is a word invented to describe nitrogen oxides.
8. Some other possible implications of the zodiacal light in history are discussed in Chapter 7.

9. *Giotto* also flew by Comet Grigg-Skjellerup in 1992, but only about half of its instrument complement was functioning after the battering it got at Comet Halley.

10. This form of cooling is called the *anti-greenhouse effect*. Such cooling is caused not only by the dust scattering away sunlight (so that the Earth absorbs less solar radiation altogether) but also through the local heating of the upper atmosphere (because the dust absorbs sunlight), resulting in more energy being radiated away from that level as infrared.

Chapter 5

1. Apart from the rocks and soil returned by the Apollo missions, the Soviet Luna program also returned several small samples of lunar soil to Earth by using robotic landers.

2. The word *terrain* is often used in connection with different areas of the lunar surface, and yet this word's etymology shows it to be inappropriate in that regard; but the correct term, *selenographical region*, seems too much of a mouthful.

3. A notable exception is Comet Encke. This was first spotted in 1786 and observed also in 1795, 1805, and 1819, since when it has been tracked on every perihelion passage (every three years and four months) through to the present. In later chapters we discuss P/Encke in more detail; its orbit is shown in Figure 3.

4. Hubert Newton (1830–1896) was interested mainly in how parabolic comets are captured into short-period orbits, a topic of research that is still hotly disputed.

5. Charles Messier (1730–1817) is best known to astronomers for his catalogue of nebulous objects (such as galaxies) in the sky. It is an ardent pursuit of many amateur observers to observe all 113 so-called Messier Objects. In fact, Messier compiled his catalogue so that these stationary nebulosities would not be confused with the moving nebulosities (comets), which were his main interest. Between 1759 and 1798, he discovered 13 comets, so he could afford to lose one to Anders Johan Lexell.

6. It is quite simple to show that the relative numbers of impacts at an angle θ vary as $(\sin\theta\cos\theta)$, θ being measured from the vertical. This distribution has a maximum value at $\theta = 45^\circ$, which is therefore the most likely impact angle. A vertical impact implies $\theta = 0^\circ$, and because $\sin 0^\circ = 0$, a vanishingly small number of impacts at near-vertical incidence is expected, to which extent Gilbert was correct: A vertical impact is very unlikely. To look at it another way, vertical impacts result only from projectiles whose extrapolated paths pass through the center of the Earth, and that is not a very big target.

7. It is interesting to note that this is broadly one of the misunderstandings of those who claim the recoil of JFK's head during the assassination was "against the laws of physics." They consider only momentum, not energy.
8. Ernst Öpik was a pioneer in many areas of the study of small bodies in the solar system. He was an Estonian who worked for many decades at the Armagh Observatory in Northern Ireland, and at the University of Maryland, doing much pioneering work. It was he who derived the first reasonable method for making estimates of the probability of a collision with a planet for specific asteroid, comet, and meteoroid orbits. He died in 1985 at the age of 91.
9. Paradoxically, Shoemaker and others in the employ of the USGS set up the Branch of Astrogeology of that organization in Flagstaff, not many miles from Meteor Crater. That institute is largely concerned with the study of impact craters. Thus the USGS has done an abrupt about-turn since the days of its support of Gilbert's explanation and the denigration of that supported by Barringer.
10. There are a number of ways of determining the ages of craters. For very old craters (more than millions of years), the age is indicated by the target rocks, which may be placed into different geological epochs. For recently formed structures such as Meteor Crater, however, the age can be measured directly through a technique known as *thermoluminescence*. Rocks emit a very weak phosphorescence when pulverized and heated, the light energy having been stored up since the last time that the thermoluminescent clock was reset in an extreme heating event, such as an impact. This technique is much used in archaeology: Although carbon-14 dating can be used for once-living things, like wood or cloth, it cannot be used for inanimate objects. Thus thermoluminescent dating is useful for ascertaining the ages of pottery, for example. A similar effect is *triboluminescence*: This is the emission of light by some crystals when they are crushed. Try crunching Wintergreen Lifesavers between your teeth in front of the mirror in a darkened room, and you'll see the effect.
11. There is no such thing as *proof* in science, only disproof. At many junctures, however, one must have confidence that an assumption is secure; for example, what physics could be done if we continually questioned whether Newton's Laws of Motion were valid? Although the association of shatter cones with impacts may not be quite as secure as Newton's Laws (and note that Newton's Law of Gravitation was assumed to hold until Einstein showed in 1916 that it was merely a good approximation), it seems unlikely at this stage that any other process capable of producing the necessary transient pressures will be found. We can therefore feel fairly secure that shatter cones indicate impact structures.
12. A typical impact produces a crater of about ten times the diameter of the impactor. Because 1-km objects in near-Earth orbits strike the Earth at a rate of about one every 100,000 years (see Chapter 2), 10-km craters are produced at about this rate.

Chapter 6

1. This population of comets, known as Kuiper-belt objects, have been suspected for some decades, but the first detections have been made only in the past few years.
2. But then again, maybe we're missing something obvious, or something beyond our range of knowledge is happening.
3. On the other hand, read the article (given at the end of this note) in which it is deduced that one of our nearest stars, Alpha Centauri, will pass close by the solar system in about 30,000 years and perturb the Oort cloud—a comet wave containing about 200,000 comets arriving in the planetary region over the following million years. See R.A.J. Matthews, The close approach of stars in the solar neighborhood. *Quarterly Journal of the Royal Astronomical Society*, volume 35, pp. 1–9 (1994).
4. Stellar/GMC relative speeds are tens to hundreds of kilometers per second, much higher than cometary velocities, so that the GMC would reach the planetary region well ahead of any cometary wave.
5. Tektites are small glassy spheres apparently produced in impacts, being mainly composed of melted target rock. They are found in extensive strewn fields in various parts of the world, particularly in North America, eastern Europe, west Africa, and Australasia, each field being the result of a specific impact event. For example, the Australasian strewn-field tektites seem to have been formed about 700,000 years ago, the other fields being much older.

Chapter 7

1. The Kreutz group of comets is a set of comets—some of them only a few tens of meters in size—which follow very similar retrograde orbits, with perihelia only just above the solar chromosphere. These are clearly the fragments produced when a much larger comet broke up because it passed too close by the Sun, although it is unclear whether this was due to tidal effects (as in the case of P/Shoemaker-Levy 9 passing too near Jupiter) or thermal stress.
2. Despite the fact that these ancient observations have time and again proven to be accurate, some will still doubt them. For example, we have observations of P/Halley from Babylonian tablets some centuries B.C., and Chinese sightings of P/Swift-Tuttle from A.D. 188 and 69 B.C., which have enabled us to state with confidence that the latter body will miss the Earth by 2 weeks in A.D. 2126. At the time of this writing, a comet (McNaught-Russell, or 1993v), which was rediscovered by two of my colleagues at the Anglo-Australian Observatory, is at its brightest in the night sky, almost reaching naked-eye magnitudes; indeed it was a naked-eye object its last time around, when the Chinese observed it in A.D. 574.

3. Just to confuse us, in the latter half of the 1980s as Chiron was approaching perihelion (that is, as solar heating increased), it began to show evidence of outgassing and other comet-like activity. Thus it is really a giant comet rather than an asteroid. If it had been found a few years later, after this cometary activity had started, it would have been named P/Kowal.
4. The student of mythology will recognize that the name was carefully chosen: Damocles was forced to sit at a banquet with a sword suspended above his head by a single hair. Similarly, asteroid 5335 Damocles has an unstable orbit that could lead to its plunging Earthward in the not-too-distant future.
5. Trojan asteroids are bodies that have the same orbital period as one of the planets, but remain about 60° in front of or behind that planet in its orbit. Jupiter has many known Trojans. Trojan orbits seem to be stable, because the bodies occupying them cannot come close to the planet in question.
6. For example, the rings of Saturn are believed to have been formed when a comet flew much too close by the planet (within a few planetary radii). Under such circumstances, it is easy to show that a large body will be broken up, based on the assumption that it behaves like a liquid; that is, that it has zero cohesive strength. This assumption is valid for most large objects (unless made of reinforced concrete)—their self-gravitation is all that holds them together. A mountain-sized rock in space tends to have zero cohesive strength: it will have fault lines or cracks through it, but once it has shattered along those lines, the smaller boulders do have significant cohesive strength. Thus Saturn's rings contain many large boulders, and even moonlets, despite the calculations based on the aforementioned assumption indicating that they should fall apart. Similarly, I live in a house made of mud-bricks (or adobe). Those bricks have very little cohesive strength and so easily crumble, but they have high compressional strength and so can easily bear a large weight.
7. One might ask, then, why P/Halley and P/Swift-Tuttle do not have diffuse streams since they have been around for many millennia. The answer is that they both have retrograde orbits and therefore only make high-speed approaches to any of the planets, these being ineffectual in altering the orbits. For P/Swift-Tuttle, the situation is acute in that its orbit is oriented almost perpendicular to the ecliptic with perihelion in the inner solar system, with the result that it never comes close to any of the giant planets, and so never suffers any major perturbations.

Chapter 8

1. In particular, there seemed little scientific logic behind the azimuth of the main axis that Lockyer decided to use, the choice of angle being critical for the date estimation.
2. Fred Hoyle is an exceptional British astrophysicist/cosmologist who has made

a large number of important scientific contributions, although he has often been controversial in his actions and his scientific beliefs. For example, he has suggested that comets bring new strains of flu and other infections to the Earth. He has also published numerous science fiction novels, and has an abiding interest in Stonehenge.

3. Anyone who has ever visited a really productive observatory or scientific laboratory will know that things tend to be haphazard and utilitarian. If it all looks neat and ordered, and worse still *comfortable*, then it is unlikely that the institution in question is at the cutting edge of science.
4. This concept of latent utility is brilliantly discussed by Stephen Jay Gould in his collection of essays entitled "Bully for Brontosaurus: Reflections in Natural History" (W.W. Norton, New York, 1991).
5. Poulat Babadzhanov and Yuri Obrubov, who are undoubtedly the world's experts on the orbital evolution of short-period meteoroid streams, work at the Institute of Astrophysics in Dushanbe in Tajikistan.
6. LDEF had a planned operational lifetime of less than a year. The *Challenger* disaster left it stranded in space for another half-decade.
7. These two showers are called the Zeta Perseids and the Beta Taurids. The star Zeta Persei is at the far southern edge of Perseus, where it meets Taurus. It has often struck me that the Beta Taurids are misnamed, as follows. Fred Whipple, after his studies of the nighttime Taurids in the 1930s, had predicted the existence of the daytime intersections with the stream. These were duly found when radars became available for meteor detection after World War II. One of the first groups to study these showers was located at Jodrell Bank, part of the University of Manchester in England. They determined the radiants and orbits of these showers, which are still used by most meteor workers in various studies, perhaps unwisely. The radiant position quoted is nowhere near the star Beta Tauri, and because this shower is part of the stream observed at its ascending node, the radiant must be south of the ecliptic, whereas Beta Tauri is to the north. On the day of peak activity (June 30) the Jodrell Bank team found a radiant that was reasonably near Beta Tauri, but they also noted that the meteor activity came from an area more than 10° across. It turns out that their mean radiant is actually closest to the star Zeta Tauri rather than Beta Tauri. I wondered whether a small transcription error has led to the shower being misnamed, misleading us for decades, but on checking their very first observations (from 1947), I found that those gave a radiant further north, closer to Beta Tauri, and the name stuck. The same thing applies to the Zeta Perseids: In fact, the radiant measured later was further south, closer to the Pleiades than Zeta Persei.
8. Examples of the consequences of the Julian calendar, which may still be seen today, are the so-called Crooked Churches of England. Many churches were

to be oriented not toward due east, as many believe, but toward the point on the horizon where the Sun would rise on the saint's day for whom the church was dedicated. A church built in honor of St. Mary would be aligned with where the Sun rose on March 25, for example. Over the course of a century or so, however, for dates according to the Julian calendar, the Sun would rise noticeably farther south than the church axis. When a new chancel was constructed, it would be aligned with the contemporary position of sunrise on the appropriate day, and so the church would become crooked. In Oxfordshire, alone, there are 81 crooked churches, a phenomenon that petered out when the Gregorian calendar was adopted. Clearly, the Roman Catholic church had more influence on the churches of England than most people realize.

9. I have slipped into Latin (per annum) to avoid having to write *year* again, and then define which year I mean.
10. However, the direction of the rising shower radiant at the Spring Equinox could be related perhaps to the orientation of many other ancient sacred structures, such as the pyramids and the great Sphinx in Egypt, which face east.
11. When the advent of meteor radars made it possible to observe meteor orbits and radiant distributions, it was found that there were three dominant directions from whence meteoroids arrived: the *apex source* (particles meeting the Earth head-on), the *Helion source* (particles with radiants near the Sun, arriving on the dayside), and the *anti-helion source* (those with radiants directly opposite the Sun, arriving on the nightside). Although each source is quite diffuse, the recognition of three main arrival directions emphasizes the recent provenance of many of the meteoroids present in this epoch. The helion and anti-helion sources derive largely from the partially dispersed Taurid Complex.
12. Alternatively, we might note that the showers due to the ascending and descending nodes of the stream produce showers with radiants split a few degrees each side of the ecliptic (that is, each side of the Sun). The positions for the Heel Stone and its presumed twin were (when the Heel Stone was upright) two or three degrees from the axis. Thus these stones could have been aligned with the shower radiants. In addition, I note that spectacular activity from the two branches would occur in different epochs centuries apart, so maybe there never was a twin stone, but rather the Heel Stone was switched in position.
13. My house is 4 kilometers from the Anglo-Australian Telescope; it cost me thousands of dollars to light-proof it in accord with the local lighting ordinances, which are designed to protect the observatory from light pollution.
14. "What needs my Shakespeare . . . that his hallow'd reliques should be hid under a Starrypointing Pyramid?" —John Milton.

Chapter 9

1. The radars used in meteor detection are unlike most people's visualization of a radar, such as a dish at the airport scanning around the sky. Meteor radars operate at lower frequencies/longer wavelengths (typically 5 to 10 meters), so that the antennas used may be wires stretched between posts or may look like TV aerials. Meteor radars typically detect meteors at a rate of one every five or ten seconds, far in excess of the visual rate of about ten per hour for someone with good eyesight.
2. In Chapter 8, we discussed why the Gregorian calendar is superior to the Julian. In 1917, Russia was still using the Julian calendar, with a result that the famous October Revolution actually occurred on November 7. This is significant in that we must take it into account when determining the date in our (Gregorian) calendar that Tunguska actually happened, that being important with respect to the discussion of the origin of the Tunguska object.
3. The pressure wave transmitted through the air has been the most important measure of the blast power. A more recent example of a similar technique is quite famous, as follows. At the time of the Trinity test in New Mexico in 1945, the first atom bomb ever exploded, the Italian-American physicist Enrico Fermi (1901–1954) made a quick and crude estimate of the energy released by scattering torn pieces of paper around him on the ground, and then seeing how far they moved when the air wave from the explosion came through his position, which was kilometers from the blast location.
4. I have earlier suggested that much research done in the former Soviet Union is ignored by westerners. One of the problems is that most anglophones do not speak or read any other language, with Russian being one of the least likely languages to be comprehended. Because the usual language of science is English, scientists whose mother tongues are other than English are disadvantaged—research papers are unlikely to be noticed unless they are published in English. Even then, it is an uphill battle. For example, two very useful papers by teams of U.S. authors on the detonation/fragmentation of small asteroids in the atmosphere, with special application to the Tunguska event, were published early in 1993 and discussed in this chapter. However, neither refers to the earlier paper by Korobejnikov and colleagues, which covers very similar ground and comes to broadly similar but distinct conclusions. The latter paper was published in Russian in early 1991, and its English translation appeared in November 1991, but nevertheless it was neglected.
5. One of the most interesting types of meteorite is that class called *carbonaceous chondrites*. These are black and generally crumbly and contain large fractions of organic chemicals. They are viewed as being some of the most primitive objects in the solar system, and as such hold important clues as to how the planets formed. Two of the most famous carbonaceous chondrites are the

Allende (Mexico) and the Murchison (Australia) meteorites, both of which fell during the 1960s. Carbonaceous chondrites are quite rare because they are made of highly friable material, which is unlikely to reach the ground intact. A few do survive atmospheric entry, however, if they arrive with low enough speeds and shallow flight angles. Of course, a large enough asteroid made of such material will hit the ground without having been significantly impeded by the atmosphere. It is believed that the 14-kilometers-diameter Lappajärvi crater in Finland was produced by a carbonaceous chondrite about 1 kilometer in diameter.

Chapter 10

1. What is described here is a *reflecting* telescope; *refracting* telescopes, which use large lenses to collect the light, are limited by the largest pieces of glass that may be manufactured without faults within them—faults would distort the images. The largest ever built have lenses about a meter in diameter, whereas it is possible to make much larger mirrors. No large refracting telescopes have been built since 1900.
2. The amount of light scattered by the atmosphere is also much less at the red end of the spectrum, making such photographs crisper. To demonstrate this, get a red and a blue filter (for example, gels used in theater footlights) and compare the scenes you see of a distant object (say, a mountain more than a few miles away) through each. The amount of scattering is wavelength-dependent. For this same reason, Earth-resources scanners and spy satellites tend to operate in the red or near-infrared part of the spectrum.
3. 719 Albert is notable in that it was numbered and named when the rules for such formalities were somewhat laxer than they are today. As a result, although currently, once an object is numbered and named, we have a good-enough knowledge of its orbit to be able to easily find it again at any time, earlier this century several asteroids became "lost." Most of these have been tracked down, however, often through the strenuous efforts of Richard West of the European Southern Observatory. In 1991, the second-last lost asteroid, 878 Mildred, was reidentified by Gareth Williams (Harvard-Smithsonian Center for Astrophysics), with the help of Rob McNaught (Anglo-Australian Observatory). In fact, 878 Mildred was discovered in 1916 from Mount Wilson Observatory in California by S.B. Nicholson and Harlow Shapley, and was named by Shapley for his new baby daughter, who was delighted when "her" asteroid was secured so many decades later; Mildred Shapley Matthews has for some years worked at the University of Arizona, editing books in the Space Science series published there. In 1995, 719 Albert remains the only lost asteroid, but there are efforts, led by Ted Bowell (Lowell observatory, Flagstaff, Arizona) to find it.

4. In fact, an asteroid with an Aten-type orbit had been discovered two decades earlier (1954 XA), but the observations were not extensive enough to secure it. Hopefully, it will be rediscovered "by accident" one day.
5. It is a fallacy that astronomers need larger telescopes to make things look bigger. The need for larger telescopes is dictated by the requirement of greater light grasp to study fainter objects. In fact, the amount of detail detectable is limited by the scintillations produced by the atmosphere. A backyard telescope with a 10-centimeter aperture has as good an angular resolution as the largest ground-based research telescope. This atmosphere-imposed limitation was one of the factors that made a space-based instrument like the Hubble Space Telescope necessary.
6. See Chapter 7, note 5. In 1990, a Mars-Trojan was found (1990 MB, now named 5261 Eureka), opening up the possibility that there might be Earth-Trojans. These would be important because they would be easily accessible, having zero velocity relative to the Earth, and hence be an economical source of raw materials for manufacturing endeavors in space.
7. In fact, this telescope is ancient—more than 70 years old—and because its right ascension drive is so worn, it cannot be used for any normal astronomical observations requiring long-term tracking.
8. In particular, the outer solar system asteroids 5145 Pholus and 1993 HA_2 are noteworthy. Although these large (bigger than 100-kilometers) objects do not come close to the Earth now, by following their orbits, by computer, into the future, it is believed likely that they could evolve onto orbits entering the inner solar system, and physical considerations indicate that they might well break up as they do so. Because each asteroid has a mass equivalent to a million 1-kilometer asteroids (recall that the present population on Earth-crossing orbits is 2,000 to 3,000 larger than that size), this would clearly have important implications for the Earth's collisional history and future. The implications of this sort of event were discussed in Chapter 7.

Chapter 11

1. Although Arthur C. Clarke writes science fiction, his work nevertheless is renowned as being the most firmly based on science fact among such writers. I have seen several physics papers published in reputable journals which discuss the plausibility of the scenarios that he invents, which rarely if ever contravene the laws of physics.
2. In the late 1980s, Andrea Milani and colleagues at the University of Pisa, Italy, also borrowed the Spaceguard name from Clarke's novel, as a title for a research program that they were undertaking. This was an appropriate name-grab: They were looking at the orbital evolution of asteroids in the inner solar

system, in particular, investigating whether the statistical calculations of impact probabilities (such as those presented in Chapter 2) are physically realistic.

3. Depending on a particle's structure, its scattering efficiency can vary between the reciprocal of the wavelength (which is termed *Mie scattering*) to the reciprocal of the fourth power of the wavelength (which is termed *Rayleigh scattering*). The sky is blue due to Rayleigh scattering by the atoms and molecules in the atmosphere.

Chapter 12

1. It has been assumed that the impactor is merely absorbed by the asteroid, thus depositing its momentum. In fact, the impact would result in a crater being formed, ejected material being rapidly thrown out in directions within the hemisphere from which the impactor arrived; these would carry away momentum in the opposite direction and therefore apply an additional "jet force," causing an even larger velocity change than the minimal value calculated here. A similar consideration shows why most people's concept of how a spectroscope works is wrong: A spectroscope is a largely evacuated glass ball similar to a large light bulb, containing a rotor comprising four diamond-shaped metallic paddles, each having one side that is polished, and one side that is painted black. The common (but incorrect) idea of why the rotor spins is that the black sides absorb light photons and are pushed backward by the photon momentum, whereas the shiny sides reflect the light and so do not absorb any momentum. In fact, the momentum change of the photons on being reflected is up to twice that which occurs when they are absorbed, so that if photon momentum were involved, the rotor would spin the opposite way. The way that a spectroscope actually works is that the black sides are heated by the absorbed light energy, and the few air molecules still within the bulb absorb some of this heat when they meet a black side, then recoiling with an enhanced speed and applying an impulse to the rotor. If the bulb were *totally* evacuated (a pure vacuum), the spectroscope would not work, except if the rotor is sufficiently friction-free and the light source very intense, in which case the rotor would spin the opposite way—due to photon momentum pressure.

2. One possibility might be to use a kinetic impulse to divert a smaller asteroid (say, 100 meters in size) in the main belt to place it on a collision course with the Earth-threatening body while the latter is near aphelion, assuming that it does have a conveniently large aphelion distance. One would therefore have available a much larger mass for diverting the dangerous asteroid. However, (1) we do not have available a knowledge of the population of smaller main-belt asteroids; (2) this mechanism would add a significant degree of complexity to the operation; (3) one would likely have to apply many impulses to

achieve an intercept course, right up until the impact because the rogue asteroid presents only a 1-kilometer target, and we would be uncertain of its position to within thousands of kilometers; and (4) if the impact by the smaller asteroid upon the larger were too energetic, the latter would be broken apart, with many of the fragments propagating on and impacting the Earth.

3. In the center of the Sun, the nuclear fusion reactions are continually converting mass into energy at a rate of more than 4 million tons per second; although this sounds like a lot, it is actually only about one part in 15,000 billion of the solar mass. This energy source will not be used up soon.

4. Indeed one could argue that the SDI (Star Wars) program brought about the end of the Cold War by bankrupting the Soviet Union: Every time the United States "upped the ante," the Soviet Union had to respond in a like manner, until it could no longer bear the burden of doing so—an effective way to win a war without anyone suffering, apart from the taxpayer.

5. Among other tasks, *Solwind* produced images of the solar corona by occluding the Sun with a disk. In such images, six sun-grazing comets were found, apparently all members of a group (called the Kreutz family—see Chapter 7, note 1) that was produced long ago when a much larger comet broke up apparently due to tidal stress—the same way that P/Shoemaker-Levy 9 broke up near Jupiter in 1992. *Solwind*'s sister craft, the *Solar Maximum Mission*, produced similar images in which another ten Kreutz family comets were discovered in 1987–89.

6. Geographos was discovered from the Palomar Observatory in 1951 during a sky survey supported by the National Geographic Society—hence its name.

7. NASA, however, was involved in helping with the space navigation. Don Yeomans of the Jet Propulsion Laboratory, in particular, was involved. The science team was led by Gene Shoemaker, even though he had just formally retired from the U.S. Geological Survey.

Glossary

albedo The fraction of incoming light (normally from the Sun) that is reflected by an object. The Earth has a mean albedo of 37%, while that of the Moon is 11%, and that of Mars is 15%; the high terrestrial albedo is partially due to the reflectivity of clouds. Most asteroids have albedos in the range 2% to 20%. Comets appear bright because their comae and tails scatter a great deal of light, although the actual nuclei have low albedos (similar to asteroids).

Amor asteroids Asteroids that have perihelion distance $q < 1.3$ AU (an arbitrary cut-off) but $q > 1.0167$ AU. The archetype, 1221 Amor, was discovered in 1932, the same year as 1862 Apollo, although in 1898 the first Amor-type (433 Eros) had been found. Although these asteroids cannot strike the Earth in the present epoch, in the longer term (generally at least some millennia), an impact is possible as the orbit of each gradually evolves under planetary perturbations (the long-distance tugs of the gravitational fields of the planets). A more rapid orbital evolution is possible if the asteroid makes a close approach to one of the planets.

aphelion The furthest point in an orbit from the Sun. The aphelion distance (symbol Q) is given by $Q = a\,(1 + e)$. The Earth has $Q = 1.0167$ AU.

Apollo asteroids Asteroids that cross the Earth's orbit (and therefore have perihelion distance $q < 1.0167$ AU and aphelion distance $Q > 0.9833$ AU) and also have periods of greater than 1 year (so that their semi-major axis $a > 1$ AU). The archetype, 1862 Apollo, was discovered in 1932.

asteroid A solid, rocky body in space, ranging in size from the larger meteoroids (5 to 10 meters) to a maximum of about 1,000 kilometers in the case of asteroid 1 Ceres, the first such body discovered. Ceres was found on the first day of the nineteenth century (January 1, 1801). Asteroids with well-determined orbits are given a permanent number, the discoverers having the privilege of suggesting a name to the International Astronomical Union's naming committee. Examples include 12 Victoria (for Queen Victoria), 1034 Mozartia, 1815 Beethoven, 2001 Einstein, and 2985 Shakespeare. Seven asteroids in a row—3350 to 3356—take the names of the seven astronauts who died in the

Challenger disaster, while another four in a row are named Lennon, McCartney, Harrison, and Starr. (Not all names are entirely serious. One astronomer named an asteroid Mr. Spock; not for the character in the *Star Trek* TV program, but for the pet cat who had been named for the TV character. Of course, they both have pointed ears). The current numbering is up to about 6,000, although many tens of thousands of asteroids have been spotted, but their orbits have not yet been accurately delineated. Such asteroids are given preliminary designations: The first asteroid found in the first half-month of the year is called (for example) 1995 AA, followed by 1995 AB, and so on. The first found on or after January 16 is 1995 BA, then 1995 BB, and so on. The letters I and Z are not used for the half-month designations (due to the chance of confusion with the numerals 1 and 2). All letters of the alphabet, except for I, are used in the second part of the designation. If more than 25 asteroids are found in any two weeks, then a subscripted numeral is used: for example, 1995 HA_1, then 1995 HB_1, and so on. The fifty-first asteroid found in the second half of April 1995 would be 1995 HA_2. The record number of asteroids given designations in any half-month was set in March 1981, the tally being 1244, and the final one being 1981 ET_{49}: This may indicate to the reader why the problem of finding possible Earth-impacting asteroids is just like the proverbial search for a needle in a haystack. As if all of this were not confusing enough, in the past, some specific survey programs have developed their own designation system: For example, in the early 1960s, a joint program of the Palomar Observatory in California and the Leiden Observatory in Holland discovered many new asteroids, and these have a numerical ordering followed by P-L, such as 6753 P-L. The word *asteroid* means "star-like," because asteroids appear similar to stars (that is, unresolved points of light) when viewed through ground-based telescopes.

asteroid belt The majority of asteroids yet discovered in the solar system inhabit orbits in a broad belt between 2 and 4 AU from the Sun (compare: Mars orbits at 1.38 to 1.66 AU, Jupiter at 4.95 to 5.45 AU).

astronomical unit The mean distance between the Sun and the Earth, about 150 million kilometers. Abbreviated as AU.

Aten asteroids Asteroids that cross at least part of the Earth's orbit but have periods of less than 1 year (thus $a < 1$ AU). These have aphelia $Q > 0.9833$ AU. The archetype, 2062 Aten, was discovered in 1976. Another asteroid, apparently of this type, was discovered more than two decades before (1954 XA), but it was only observed for six days and is now lost.

biosphere The region of the Earth within which living organisms are found. Although this might be thought to include only the surface and the oceans, in recent years it has been found that bacteria exist in colonies deep within crustal rocks, extending the biosphere beyond previous concepts.

bolide Another name for a brilliant meteor, especially one that appears to explode or detonate.

cometary apparition Each passage of a comet through its perihelion point is called an *apparition*. The first comet observed in each year (either a newly discovered comet, or a rediscovery/recovery of a periodic comet) is denoted by the lowercase letter a, the next by b, and so on; for example, 1994a, 1994b, 1994c, and so on. The letter must be lowercase because uppercase letters are used for supernovae. If more than 26 comets are found in any year, as has been the case since 1987, a subscripted number is used: for example, $1994a_1$, $1994b_1$, and so on. It has not yet been necessary to graduate to a subscript 2 (that is, in no year have more than 52 comets been found). When the comet discoveries for a year are complete, they are given a new permanent designation with a roman numeral, which is an ordering in terms of the time of perihelion passage within that year. Thus the first comet passing perihelion in 1994 is 1994 I, the next 1994 II, and so on. For example P/Halley may be referred to as 1835 III, 1910 II, or 1986 III (plus many earlier apparitions). This labeling of comets is due to be changed beginning in 1995, with a new system making use of the half-month letters as used for asteroids.

comet names Apart from getting alphanumeric designations (which are the handles used most frequently by professional astronomers), a comet also gets the name of its discoverer. If more than one person discovers a comet before an announcement is made through the IAU, then up to three individual names may be allotted to any comet, priority resting with the times at which the reports are made to the Minor Planet Center at the Harvard-Smithsonian Center for Astrophysics in Cambridge, Massachusetts. An example is Comet West-Kohoutek-Ikemura (1987 XV). Occasionally, an instrument name may be allocated to a comet when a team has used that particular instrument in the discovery; for example Comet IRAS-Iraki-Alcock (1983 VII), IRAS being the Infra-Red Astronomy Satellite, which functioned in 1983, or Comet Spacewatch (1990 XXIX). There are several comets with the same name (for example, Comet Bradfield, Comet Shoemaker-Levy) because the discoverers have had multiple successes—thus the usefulness of the alphanumeric designations. The rule that a comet is named for the discoverer is not obeyed in the cases of some comets found in the past. For example, Sir Edmond Halley did not discover P/Halley; it bears his name because he predicted its return in 1759. Also, common usage prevents the renaming of a comet that is later recognized as having been seen on a previous apparition; for example, P/Swift-Tuttle retains that name for the American astronomers Lewis Swift and Horace Tuttle who discovered it in 1862, despite the fact that it has recently been realized that the Jesuit missionary Ignatius Kegler saw it from China in 1737.

comets Substantially sized planet-crossing objects mainly composed of ices, but with a significant fraction made of refractory material. Although, so far, humankind has sent spacecraft to only one comet (P/Halley), the data from those spacecraft and from telescopic observations support the idea that 50% to 80% of the mass of a comet is water ice. Comets are relatively easily seen in the sky because, as they enter the inner solar system from elongated orbits, their volatile component begins to evaporate, producing a coma (a cloud of vapor about the solid nucleus, this nebulous envelope being 10,000 to 100,000 kilometers in diameter) and also a tail, which may be millions of kilometers long: the word *comet* is derived from the ancient Greek for "long-haired." In fact, comets generally have two distinct tails: an ion tail, made up of charged atoms produced when molecules in the coma are split apart and then ionized by the solar ultraviolet and x-ray flux, and a dust tail made up of dust grains and meteoroids originating in the refractory component of the nucleus, caught up in the gaseous outflow into the coma. The ion tail does *not* show the direction of motion of the comet, but is aligned with the flow of the solar wind, a continuous stream of charged particles moving outward from the Sun; the ion tail, which often appears slightly bluish, therefore points in the anti-helion direction. The dust tail forms a fan-shape between the ion tail and the anti-direction of the comet's motion. Dust particles are affected, depending on particle size, by the solar wind and solar radiation pressure, which has the effect of increasing the size of their heliocentric orbits. The dust tail usually appears pinkish-red. Because the detectable tails of comets may be more than 1 AU long, comets as such might be claimed to be the largest objects in the solar system. As the heliocentric distance of a comet decreases, the more volatile ices (methane, ammonia, carbon monoxide, carbon dioxide, methanol, etc.) begin to evaporate from its surface. At 3 AU from the Sun, the heating is sufficient for water to start to vaporize, so that the coma becomes more extensive and the comet brightens. The sizes of cometary nuclei are uncertain because when they are in the inner solar system, they are obscured by the comae that surround them. It is believed that most observed comets have nuclei about 1 kilometer in size. The brightness of a comet depends on its distance from the Sun, its distance from the Earth, the size of its nucleus, and how much of the nucleus is actively evaporating (because comets apparently form insulating layers of refractory chemicals and heavy organics—like tar—over their more volatile interiors). P/Halley was shaped like a potato (some say a peanut) with its longest axis about 15 kilometers in extent. P/Swift-Tuttle is thought to be of the same size or larger, while P/Encke is about 5 kilometers across. On each passage through the inner solar system, a comet may lose a meter or so of its material, as an average across the whole surface, as mass lost into its coma, and from that source into its ion and dust tails; thus it is not surprising that P/Halley and P/Swift-Tuttle have been observed for over 2 millennia and may well have been in their present orbits for upward of 20,000 to 50,000 years. Long-period comets, especially those on their first trip through the inner solar system from the Oort Cloud, tend to be much brighter than comets that have passed through repeatedly, presumably because the pristine surfaces of "new" comets are not yet choked with refractories and heavy organics. The brightest (and hence largest) comets ever observed have been estimated to be 50 to 100 kilometers in size; of particular note in historical times was Comet Sarabat in 1729, which was exceptionally bright despite the fact that it never came closer than 4 AU from the Sun, so there was no substantial water evaporation. There is no known lower size limit, although clearly, if exposed volatile material boils off at a rate of a meter or so per perihelion passage, a 20- or 50-meter comet cannot last for long. If a comet has some remnant refractories left over after complete devolatilization, it would be counted as an asteroid (because that is the way in which it would appear, telescopically), even though in reality it is an extinct comet. A comet might equally well form a complete insulating crust, appearing to be an asteroid even though it is a moribund comet, a period of dormancy possibly being broken when some capricious meteoroid impact breaks through the crust and exposes fresh volatile material.

earth-approaching asteroid, near-earth asteroid Any of the Aten, Apollo, and Amor classes.

earth-crossing asteroid One of the Aten and Apollo classes.

eccentricity A measure of the deviation from circularity of an orbit. A circle has eccentricity 0, a parabola has eccentricity 1, and a hyperbola has eccentricity greater than 1. See, for example, the comet and asteroid orbits in Figures 1 through 4. Abbreviated as the letter *e*.

ecliptic The plane of the Earth's orbit. All of the planets have orbital planes close to that of the Earth's, with the largest inclination being that of Pluto ($i = 17°$).

fireball An exceptionally bright meteor.

geocentric orbit An orbit about the Earth.

heliocentric orbit An orbit about the Sun. The word stem *heli-* implies the Sun; for example, the chemically inert element helium derives its name from the fact that it was first identified in the solar atmosphere, although it is now known to be abundant on Earth (for example, in natural gas).

ices A generic term used by astronomers to cover the four basic types of solid material found in space which we on Earth are more accustomed to experiencing as gases or liquids: water ice, methane, carbon dioxide, and ammonia.

inclination The angle between a particular orbital plane and that of the Earth (the ecliptic). Its symbol is the letter *i*.

inner solar system The region closer to the Sun than to Jupiter (that is, heliocentric distances less than about 5 AU).

intermediate-period comet Also know as *Halley-type comets*, these have periods in the range 20 to 200 years. The two comets for which the longest time bases of observations are available are Comet Halley (records stretch back continuously to 240 B.C.) and Comet Swift-Tuttle (observed in A.D. 1992, 1862, 1737, and 188, and 69 B.C.). These two comets have orbital periods of 76 years and about 130 years, respectively.

International Astronomical Union (IAU) The IAU is the world governing body of astronomy, with committees having responsibility for, among other things, the naming of celestial objects (such as asteroids and comets and planetary surface features).

interplanetary dust (IPD) A solid particle too small to be termed a meteoroid (because it would not burn up in the atmosphere as a meteor, instead being gradually decelerated and remaining largely intact). Sizes are less than 0.1 millimeter. IPD originates from particles released by asteroids and comets; most small (size: <1 centimeter) meteoroids have their lifetimes limited by catastrophic collisions with IPD grains, producing more IPD in their breakups.

Jovian planet, giant planet, gas giant, outer planet These four equivalent terms are used to refer to any of the large planets: Jupiter, Saturn, Uranus, and Neptune.

Kuiper belt A theoretical construct put forward by the Dutch-American astronomer Gerard Kuiper in 1951, largely on the basis that one would not expect the solar system to suddenly stop at the outermost planet if the planets did indeed form from many smaller planetoids colliding with each other. Kuiper reasoned that there should be many planetoids left over in the region 40 to 100 AU from the Sun, which were never incorporated into the planets. Experimental back-up for this idea has come in the few years since 1992 with the discovery of more than a dozen 100- to 400-kilometer objects at heliocentric distances of about 40 AU. These are currently classified as asteroids, although their true nature is unclear. The Kuiper belt, rather than the Oort cloud, is favored by many as the likely source of most short-period comets. Inclusion of the word *belt* derives from the fact that the distribution of comets is thought to be largely constrained to be near the plane of the planetary orbits.

major planet One of the nine largest objects on heliocentric orbits: the terrestrial planets Mercury, Venus, Earth, and Mars; the gas giant or jovian planets Jupiter, Saturn, Uranus, and Neptune; plus Pluto. Pluto (being only about 2,300 kilometers across) is in some ways best considered as being a largish comet or minor planet rather than a major planet as such.

meteor Synonym for *shooting star*. This term relates to the phenomena produced by a meteoroid as it enters the atmosphere at a speed between 11 and 73 kilometers per second (between 40,000 and 260,000 kilometers per hour); it does not refer to any solid body, so the way in which it is used most often is incorrect. The streak of light is the meteor, that streak being termed a *trail*. If the glowing streak remains for some time, it is called a *train*. Meteoroids also produce a train of ionized (charged) particles that may be detected by radar, allowing the terrestrial influx of meteoroids down to about 0.1 millimeter in size to be determined. Typical meteor altitudes are 60 to 120 kilometers (40 to 80 miles).

meteorite If a solid fragment of a meteoroid survives entry through the atmosphere, it is called a meteorite. Usually only a small fraction of the original object reaches the ground intact, depending on the size, speed, composition, density, and entry angle of the meteoroid. Very small meteoroids and interplanetary dust particles may survive entry, especially if they are largely refractory and have low entry speeds (below 15 kilometers a second) and shallow entry angles, because they lose their energy radiatively as they enter the upper atmosphere and heat up, rather than melting and then ablating (burning up). This is a problem of mass (and hence total energy) against surface area (and hence area from which heat can be radiated away). The small meteorites produced in this way—it has been estimated that one in a thousand dust particles in a typical house may be such a meteorite—are called micrometeorites. The larger meteorites come in a variety of classes, such as iron and stoney meteorites. Meteoroids made of nickel-iron are the most likely to survive atmospheric transit and produce a meteorite, and carbonaceous meteoroids (composed largely of heavy organic materials and having an appearance akin to coal) are the least likely to reach the ground intact: This is a pity because, scientifically speaking, carbonaceous meteorites are the most interesting and appear to be tied up with the origin of the organic chemicals on Earth which are essential to life. They also hold clues to the origin of the solar system itself because they are the oldest known unaltered objects.

meteoroid A body in space, ranging in size from the upper limit of interplanetary dust (about 0.1 millimeter) to the smaller asteroids, which may be telescopically discovered in space (about 5 to 10 meters across for very near Earth passes). Meteoroids are thought to be mainly rocky bodies, although there is also evidence for a substantial organic/carbonaceous component. Most meteoroids are apparently spawned by comets as they pass through the inner solar system, the solar heating evaporating some of the ice and other volatile material of which the cometary nucleus is largely composed, with some of the solid lumps on the surface of the nucleus being carried off with the general flow. There is also evidence for some meteoroids originating from Earth-crossing asteroids, although this may simply indicate that these were previously active comets and have since lost all of their volatile material. Meteoroids quickly lose all of their volatiles (ice, methane, carbon dioxide, ammonia, and organic chemicals with less than about 20 carbon atoms per molecule) because these are thermally unstable; they were, previous to meteoroid release, shielded within the cometary nucleus. Some larger meteoroids may also be derived from the asteroid belt, either having been small bodies throughout their residence lifetime there (since the solar system formed 4.5 billion years ago) or else chips knocked off of the surface of larger bodies in inter-asteroid collisions. Meteoroids are investigated through (1) the meteors they produce as they enter the Earth's atmosphere; (2) the meteorites that occasionally survive atmospheric entry; and (3) the impact craters they have formed on the Moon and other solar system bodies with zero or low-density atmospheres—Mercury, Phobos, and Deimos, and to a certain extent Mars, the asteroids of which we have detailed images, and the moons of the outer planets Jupiter, Saturn, Uranus, and Neptune.

minor planet A synonym for *asteroid. Minor planet*, however, is the term officially favored by the IAU.

near-earth object Any asteroid or comet that can come close to the Earth's orbit (that is, $q < 1.3$ AU).

Oort cloud A hypothetical reservoir of comets at the periphery of the solar system (that is, out to about a quarter of the distance to the nearest star). This idea was suggested by Dutch astronomer Jan Oort, in 1950, on the basis of the groupings in orbital energy that he noted among the long-period comets that appeared to be on their first trip through the planetary region from a great distance. Typical Oort cloud comets have semi-major axes on the order of 50,000 AU, implying orbital periods of about 10 million years. Comets are believed to be redirected occasionally onto orbits with small perihelion distances (making them observable) due to various perturbations, including passing stars or giant clouds of interstellar molecules. Some small fraction of the observable comets thus produced may pass much too close to one of the planets, in particular Jupiter, and be captured into a much smaller orbit. Paradoxically, it is believed that Oort cloud comets formed in the Uranus-Neptune region of the solar system and were thrown out onto their gigantic orbits by those planets soon after the planets themselves condensed and coagulated. The Oort cloud is believed to be spherical in shape, this belief being supported by the inclination distribution of long-period comets.

orbit The elliptical (curved) path followed by an object under the dominant gravitational influence of another more massive object. This larger object occupies one of the two foci of the orbit. In the case of a circular orbit, the two foci are coincidental, and the orbit is said to have zero eccentricity. The planets all have low-eccentricity, near-circular heliocentric orbits, Pluto having the largest eccentricity (about 0.25); the Earth's eccentricity is 0.0167. A more elongated orbit has a larger eccentricity but is still elliptical until the eccentricity reaches a value of 1. At this value, the orbit is said to be parabolic, and any object on such an orbit is on the verge of being gravitationally unbound to the object at the focus. Eccentricities greater than 1 represent unbound, hyperbolic orbits. Comets arriving from the Oort Cloud are near-parabolic, but no comet ever observed is known with certainty to have been on a hyperbolic heliocentric orbit (that is, not bound to the Sun and thus not a member of the solar system, having come from interstellar space).

orbital period The time taken to complete one orbit.

organics; organic chemicals Molecules that are based on carbon and are essential to life as we know it. Organic molecules with 15 or more atoms are known, from radio astronomical observations, to exist in interstellar space. They are also known to occur in some meteorites, in particular carbonaceous chondrites. Radio observations of comets had previously shown that there are organics, or at least their decay products, in cometary comae, and the various spacecraft missions to P/Halley in 1986 turned up abundant evidence that organic chemicals are plentiful in that comet. It has become customary to speak of CHON chemicals; that is, molecules containing some selection of carbon, hydrogen, oxygen, and nitrogen, hence the acronym. The actual fraction of P/Halley's mass that is organic may be up to 10%. Examples of common organics on Earth include methanol (CH_3OH), ethanol (C_2H_5OH), formic acid (HCOOH, the venom that an ant injects when it bites you), and the paraffin series starting with methane (CH_4), ethane (C_2H_6), butane (C_3H_8), propane (C_4H_{10}), and pentane (C_5H_{12}) and going on through the familiar octane (C_8H_{18}) to higher carbon numbers which become thicker than the gases and light liquids that precede them. There are other similar series, such as the alkenes (methylene, ethylene, etc.). In general, organics with carbon numbers greater than 20 or 25 are quite viscous and relatively inert. Such compounds are tarlike and are often generically termed *kerogens*. A comet with a plentiful supply of organics will preferentially lose the lighter molecules because these are more volatile, so it is possible that comets form insulating crusts of refractory materials held together with a kerogen glue. This is in accordance with the appearance of the nucleus of P/Halley, which was dark, having an albedo of less than 4%.

parabolic/long-period comet A comet with an orbital period greater than 200 years. This is a useful, but arbitrary, limit, because the vast majority of comets have been observed and recorded since the year A.D. 1800. Most long-period comets have periods much longer than 200 years. Only a few long-period comets have ever been observed to return—for example, Comet McNaught-Russell (1993v), which was last seen by the Chinese in A.D. 574. The total number of long-period comets for which reasonably accurate orbits have been determined through 1993 is 681, of which 411 had Earth-crossing orbits ($q <$ 1.0167 AU); however, many more have crude observations recorded in ancient records, in particular in the cultures of Babylon, China, and Japan. The term *parabolic comet* is often used in place of *long-period*, because these comets tend to be close to the parabolic limit (that is, the limit of being gravitationally bound to the Sun).

perihelion The closest point in an orbit to the Sun. The perihelion distance (symbol q) is given by $q = a\ (1–e)$. The Earth has q = 0.9833 AU.

planetary region The region of the solar system that is closer to the Sun than the outermost planet. Because Pluto has aphelion at about 49 AU, nominally the planetary region stretches out to about 50 AU, although (especially because Pluto is such an oddball) it would be more usual to take Neptune's orbit at 30 AU to be the limit of the planetary region.

prograde orbit An orbit going around the Sun in the same direction as the planets (counter-clockwise as viewed from the north—that is, perpendicularly above the ecliptic—and thus inclination $0° < i < 90°$).

refractories Materials that are stable under heating and are not easily liquefied or vaporized. In most cases, in this context, rocky-type material is implied—these may also be termed *silicates*.

retrograde orbit An orbit going around the Sun in the direction opposite to that of the planets (inclination $90° < i < 180°$).

selenologist A geologist who studies the Moon.

semi-major axis A measure of the size of an orbit: One half of the length of the long axis of an elliptical orbit. This is a useful quantity because, when expressed in AU, the semi-major axis (abbreviated as a) renders the orbital period in years if it is raised to the power 1.5 (that is, cubed and then square-rooted). Thus Jupiter, with a = 5.203 AU, has a period of 11.87 years, while Venus, with a = 0.7233 AU, has a period of 0.615 years (225 days).

short-period comet The old classification system divided comets into long-period (>200 years) and short-period (<200 years) groups. Currently it is more common to classify comets as being short-period if their orbital periods are less than 20 years. These are also known as "Jupiter-family comets." Individual periodic comets (both short-period and intermediate-period) are sometimes denoted with the shorthand notation P/Halley or P/Encke, for example, meaning "Periodic Comet Halley," and so on. P/Encke has the shortest orbital period of any known comet, taking 3 years and 4 months to complete each circuit of the Sun (see Figure 3). The total number of comets with periods below 200 years, which have been observed through 1993, is 174, of which only 26 have Earth-crossing orbits in the present epoch. P/Halley, P/Swift-Tuttle, and P/Encke are three of these.

solar system The complex of planets, moons, comets, asteroids, meteoroids, and interplanetary dust that is bound to the Sun by gravity.

Spaceguard Named for the fictitious program described in the first chapter of the science fiction novel *Rendezvous with Rama*, by Arthur C. Clarke, Spaceguard is the search program recommended by NASA's Near-Earth-Object Detection committee in 1992. It involves a network of at least six dedicated, purpose-built telescopes spread worldwide and coordinated from a central nexus.

terrestrial planet; inner planet One of the four planets Mercury, Venus, Earth, and Mars.

volatiles Chemicals that are physically unstable when heated, quickly attaining a gaseous state. The most abundant volatiles in comets, according to data for those objects studied to date, are water, carbon monoxide, and methanol.

zodiacal light A pyramid-shaped region of diffuse light concentrated around the ecliptic, which is seen toward the western horizon after sunset and toward the eastern horizon before sunrise. This phenomenon is caused by the scattering of sunlight by interplanetary dust particles, chiefly in the 0.01- to 0.1-millimeter range, which are arrayed in a distribution centered on the plane of the planets' orbits.

Bibliography

This bibliography lists various books, articles, and scientific papers that the reader might wish to consult to obtain more details on specific subjects. The subjects are gathered loosely by the chapter in which the subject first appears. Most entries are followed by a brief description of their content.

Chapter 1, and publications of general significance

Alvarez, L.W., W. Alvarez, F. Asaro, and H.V. Michel, "Extraterrestrial cause for the Cretaceous-Tertiary extinction," *Science*, volume 208, pp. 1095–1108 (1980). The epoch-making paper in which the evidence for an enormous impact at the time of the K/T boundary, depositing an anomalous layer of iridium, was unveiled.

Alvarez, W., "Interdisciplinary aspects of research on impacts and mass extinctions: A personal view," *Geological Society of America, Special Paper* 247, pp. 93–98 (1990). A superb discussion of why it is important to have scientists from many disciplines investigating impact phenomena on the Earth.

Chapman, C.R., and D. Morrison, *Cosmic Catastrophes*, Plenum Press, New York (1989). A discussion of the various extraterrestrial threats to one's life, and a comparison of these with other causes of accidental death.

Chapman, C.R., and D. Morrison, "Impacts on the Earth by asteroids and comets: Assessing the Hazard," *Nature*, volume 367, pp. 33–40 (1994). A broad discussion of the hazard posed to humankind by cosmic impacts.

Erickson, J., *Target Earth: Asteroid Collisions Past and Future*, TAB Books, Blue Ridge Summit, Pennsylvania (1991). A general account of terrestrial impacts.

Gehrels, T., (editor), *Hazards due to Comets and Asteroids*, University of Arizona Press, Tucson (1994). A massive collection of review chapters by the world's experts on this topic. Some specific chapters are referred to later in this Bibliography.

Hoyt, W.G., *Coon Mountain Controversies*, University of Arizona Press, Tucson (1990). An excellent, detailed account of the history of the debate about the origin of Meteor Crater in particular and terrestrial impact craters in general.

Marvin, U.B., "Impact and its revolutionary implications for geology," *Geological Society of America, Special Paper* 247, pp. 147–154 (1990). This paper discusses the history of the impact concept and how its gradual acceptance now is changing the practice of geology. Marvin also provides a sketch of the circumstances of the eventual recognition of Meteor Crater as being of impact origin.

Morrison, D. (editor), *Report of the NASA Near-Earth-Object Detection Workshop: The Spaceguard Survey*, NASA/Jet Propulsion Laboratory, Pasadena, California (1992). This report, discussed in more detail in Chapter 11, reviews the impact hazard before addressing how it should be tackled.

Mr. Statistics column, *Fortune* magazine, June 1, 1992. An economic analysis of whether an asteroid/comet search program should be carried out.

Sharpton, V.L., and P.D. Ward (editors), "Global catastrophes in Earth history," *Geological Society of America, Special Paper* 247 (1990). A volume containing many important papers discussing how the global environment has been affected in the past by impacts.

Wilhelms, D.E., *To a Rocky Moon*, University of Arizona Press, Tucson (1993). An authoritative account of the exploration of the Moon to date, including terrestrial as well as lunar cratering.

Yeomans, D.K., *Comets: A Chronological History of Observation, Science, Myth and Folklore*, John Wiley and Sons, New York (1991). An extensive volume discussing all aspects of comets.

Chapter 2

Bailey, M.E., "Comet craters versus asteroid craters," *Advances in Space Research*, volume 11(6), pp. 43–60 (1991). An analysis of the relative importance of comets and asteroids as terrestrial impactors.

Jones, H.D., "Halley and comet impacts", *Journal of the British Astronomical Association*, volume 98, p. 339 (1988). A brief letter describing Sir Edmond Halley's suggestions of the role of cometary impacts in Earth history.

Moore, P. "How the lunar craters weren't formed," *Journal of the British Astronomical Association*, volume 95, pp. 154–155 (1985). A humorous account of the various bizarre suggestions made over the years about how the craters of the Moon were formed.

Watson, F., *Between the Planets*, Blakiston Co., Philadelphia (1941); and Baldwin, R., *The Face of the Moon*, University of Chicago Press (1949). Two pioneering texts that correctly deduced that massive impacts on the Earth must occur with a worrying frequency.

Weissman, P.R., "The cometary impactor flux at the Earth," *Geological Society of America, Special Paper* 247, pp. 171–180 (1990). A calculation of the frequency of comet impacts on the Earth.

Chapter 3

Hills, J.G., and M.P. Goda, "The fragmentation of small asteroids in the atmosphere," *Astronomical Journal*, volume 105, pp. 1114–1144 (1993). A detailed analysis of the physics involved in the atmospheric entry of bolides, including tsunami generation.

Spratt, C., and S. Stephens, "Against all odds: Meteorites that have struck home," *Mercury*, volume 21, pp. 50–56 (1992). This article lists and describes structural and personal damage done over the years by meteorites.

Chapter 4

Beatty, J.K., "Killer crater in the Yucatán?" *Sky and Telescope*, July 1991, pp. 38–40. Describes the discovery of Chicxulub impact crater in the Yucatán, which has been firmly linked to the K/T mass extinction.

Bronshten, V.A., *Physics of Meteoric Phenomena*, Reidel, Dordrecht, Holland (1983). The best reference text covering the ablation of small meteoroids.

Gallant, R.A., "Journey to Tunguska," *Sky and Telescope*, June 1994, pp. 38–43. An interesting contemporary account of a trip to the Siberian taiga where the forests were blasted flat; includes information on eyewitness accounts of the event.

Krinov, E.L., *Giant Meteorites*, Pergamon Press, Oxford (1966). Includes eye-witness accounts of the Tunguska and Sikhote-Alin events of this century, and details of large impacts leaving craters.

Sagan, C., and R.P. Turco, *A Path Where No Man Thought*, Random House, New York (1990). An extensive book discussing nuclear winter.

Toon, O.B., K. Zahnle, R.P. Turco, and C. Covey, "Environmental perturbations caused by impacts," pp. 791–826, in *Hazards Due to Comets and Asteroids*, edited by T. Gehrels, University of Arizona Press, Tucson (1994). A review of how impacts of different magnitudes may affect the environment.

Turco, R.P., O.B. Toon, T.P. Ackerman, J.B. Pollack, and C. Sagan, "Nuclear winter: Physics and physical mechanisms," *Annual Review of Earth and Planetary Sciences*, volume 19, pp. 383–422 (1991). Reviews the concept of global cooling caused by the aftermath of a major nuclear exchange.

Wolbach, W.S., I. Gilmour, E. Anders, C.J. Orth, and R.R. Brooks, "Global fire at the Cretaceous-Tertiary boundary," *Nature*, volume 334, pp. 665–669 (1988).

This is a technical review of this team's work on the fires that occurred at the time of the K/T boundary event, along with related work.

Zahnle, K., "Atmospheric chemistry by large impacts," *Geological Society of America, Special Paper* 247, pp. 271–288 (1990). An analysis of the immediate effects on the atmosphere (and the surface below) of a massive impact; particularly useful for the analysis of the heating effects of the reentering rocks thrown out from the impact zone, which would ignite the global fires.

Chapter 5

Alvarez, L.W., "A physicist examines the Kennedy assassination film," *American Journal of Physics*, volume 44, pp. 813–827 (1976). The source of the comments about Alvarez's investigation of the JFK assassination.

Bronshten, V.A., "Brilliant meteor of June 1, 1937 and the Simuna crater," *Solar System Research*, volume 25, pp. 80–83 (1991). The source of information about this recent crater-forming event.

Clube, S.V.M., "The dynamics of Armageddon," *Speculations of Science and Technology*, volume 11, pp. 255–264 (1988); and Bailey, M.E., S.V.M. Clube, and W.M. Napier, *The Origin of Comets*, Pergamon, Oxford (1990). Background on the catastrophism versus uniformitarianism debate in the nineteenth century.

Gould, S.J., *Bully for Brontosaurus: Reflections in Natural History*, Norton, New York (1991). Gould details the "debate" between Huxley and Wilberforce in 1860, which in fact was not a formal debate: The events of that day have since been much misrepresented.

Grieve, R.A.F., "Impact cratering on the Earth," *Scientific American*, volume 262, pp. 44–51 (April 1990); Grieve, R.A.F., "Terrestrial impact structures," *Annual Review of Earth and Planetary Sciences*, volume 15, pp. 245–270 (1987); and Grieve, R.A.F., "Terrestrial impact: The record in the rocks," *Meteoritics*, volume 26, pp. 175–194 (1991). Three papers detailing what we know about terrestrial impact craters and how they are formed. Grieve is the unofficial "keeper of the records" on such structures.

Mark, K., *Meteorite Craters*, University of Arizona Press, Tucson (1987). An extensive study of terrestrial impact craters.

Melosh, H.J., *Impact Cratering: A Geologic Process*, Oxford University Press, Oxford and New York (1989). Excellent monograph detailing crater formation and its significance on the Earth.

Shoemaker, E.M., "Asteroid and comet bombardment of the Earth," *Annual Review of Earth and Planetary Sciences*, volume 11, pp. 464–494 (1983). A post-Apollo analysis of the lunar cratering record and what it implies for the terrestrial cratering rate.

Chapter 6

De Laubenfels, M.W., "Dinosaur extinction: One more hypothesis," *Journal of Paleontology*, volume 30, pp. 207–212 (1956). The first suggestion that cosmic impacts could have killed the dinosaurs.

Eldredge, N., *The Miner's Canary*, Prentice-Hall, New York (1991). A discussion of mass extinctions and the paleontological evidence, from one of the advancers (with S.J. Gould) of the idea of "punctuated equilibria" in evolution.

Goldsmith, D., *Nemesis: The Death Star and Other Theories of Mass Extinction*, Walker, New York (1986). The title is self-descriptive.

Gould, S.J., *Wonderful Life*, W.W. Norton, New York (1989). A magnificent description of the events surrounding the explosion of life on the Earth, with additional discussion of the ways of evolution and the influences of various factors such as impacts.

Muller, R., *Nemesis: The Death Star*, Heinemann, New York (1988). A popular-level account of the identification of a periodicity in cratering, and the thinking behind the "Nemesis" hypothesis for the origin of that periodicity.

Napier, W.M., "Terrestrial catastrophism and galactic cycles," pp. 133–167, in *Catastrophes and Evolution: Astronomical Foundations*, edited by S.V.M. Clube, Cambridge University Press, Cambridge, U.K. (1989). A strong discussion of the arguments of the underlying periodicity in cratering and other geological phenomena's being close to 15 million years rather than 26 to 30.

Napier, W.M., and S.V.M. Clube, "A theory of terrestrial catastrophism," *Nature*, volume 282, pp. 455–459 (1979). The first paper to link the movement of the solar system about the galaxy with impacts, and hence a cycle of cratering linked to geological boundaries.

Nininger, H.H., "Cataclysm and evolution," *Popular Astronomy*, volume 50, pp. 270–272 (1942). The first modern publication to recognize that massive impacts might be the cause of geological boundary events, and so be important in the evolution of life on Earth.

Perlmutter, S., R.A. Muller, C.R. Pennypacker, C.K. Smith, L.P. Wang, S. White, and H.S. Yang, "A search for Nemesis: Current status and review," *Geological Society of America, Special Paper* 247, pp. 87–91 (1990). A description of the Nemesis theory and the search for that hypothetical solar companion star.

Rampino, M.R., "Gaia versus Shiva: Cosmic effects on the long-term evolution of the terrestrial biosphere," pp. 382–390, in *Scientists on Gaia*, edited by S.H. Schneider and P.J. Boston, MIT Press, Cambridge, Massachusetts (1992). The title is self-explanatory.

Rampino, M.R., and B.M. Haggerty, "Extraterrestrial impacts and mass extinctions," pp. 827–857, in *Hazards due to Comets and Asteroids*, edited by T. Gehrels, University of Arizona Press, Tucson (1994). An up-to-date review of the evidence linking many mass extinction events—not just the disappearance of the dinosaurs—to large impacts.

Rampino, M.R., and K. Caldeira, "Major episodes of geologic change: Correlations, time structure and possible causes," *Earth and Planetary Science Letters*, volume 114, pp. 215–227 (1993). A research paper setting out the identification of periodicities in the geologic record from the past 260 million years.

Raup, D.M., *The Nemesis Affair: A Story of the Death of the Dinosaurs and the Ways of Science*, W.W. Norton, New York (1986). A discussion of the early identification of periodicities in the cratering record, the three suggested explanations for the periodicity, and the implications for the evolution of life.

Raup, D.M., *Extinction: Bad Genes or Bad Luck?*, W.W. Norton, New York (1991). Raup suggests that all mass extinctions may be due to large impact events. Excellent description of the paleontological evidence.

Seyfert, C.K., and L.A. Sirkin, *Earth History and Plate Tectonics*, Harper and Row, New York (1979). The first suggestion that impact craters on the Earth may be produced periodically.

Stothers, R.B., "Impacts and tectonism in Earth and Moon history of the past 3800 million years," *Earth, Moon and Planets*, volume 58, pp. 145–152 (1992). A description of the six distinct episodes of phenomenal terrestrial and lunar cratering that seem to have occurred in the past 3.8 billion years.

Urey, H.C., "Cometary collisions and geological periods," *Nature*, volume 242, pp. 32–33 (1973). The first paper to describe physical evidence linking big impacts with geological boundary events.

Yabushita, S., "Periodicity in the crater formation rate and implications for astronomical modeling," *Celestial Mechanics and Dynamical Astronomy*, volume 54, pp. 161–178 (1992). An advanced discussion of terrestrial crater periodicity and its implications.

Chapter 7

Asher, D.J., S.V.M. Clube, W.M. Napier, and D.I. Steel, "Coherent catastrophism," *Vistas in Astronomy*, volume 38, pp. 1–27 (1994). A technical description of the concept of coherent catastrophism.

Bailey, M.E., S.V.M. Clube, G. Hahn., W.M. Napier, and G.B. Valsecchi, "Hazards due to giant comets: Climate and short-term catastrophism," pp. 479–533, in *Hazards Due to Comets and Asteroids*, edited by T. Gehrels, University of Arizona Press, Tucson (1994). An extensive discussion of the hazard posed to humankind by giant comets, and in particular their influence on the climate.

Ceplecha, Z., "Influx of interplanetary bodies onto Earth," *Astronomy and Astrophysics*, volume 263, pp. 361–366 (1992). Deduction of the terrestrial mass influx for all particles from dust to massive asteroids and comets, making use of the important new data from the Spacewatch program.

Garwin, R.L., pp. 203–209, in *Discovering Alvarez*, edited by W.P. Trower, University of Chicago Press, Chicago (1987). The source of the quote concerning the work of Luis Alvarez.

Love, S.G., and D.E. Brownlee, "A direct measurement of the terrestrial mass accretion rate of cosmic dust," *Science*, volume 262, pp. 550–553 (1993). A paper detailing the most recent, and best, determination of the influx of dust and small meteoroids to the Earth.

Oberst, J., and Y. Nakamura, "A search for clustering among the meteoroid impacts detected by the Apollo Lunar Seismic Network," *Icarus*, volume 91, pp. 315–325 (1991). A scientific paper detailing the results from the seismographs left on the Moon, providing information on impacts by meteoroids over several years.

Pendleton, Y.J., and D.P. Cruikshank, "Life from the stars," *Sky and Telescope*, March 1994, pp. 36–42. A convenient source showing electron microscope images of interplanetary dust particles collected in the stratosphere.

Rabinowitz, D., "The size distribution of the Earth-approaching asteroids," *Astrophysical Journal*, volume 407, pp. 412–427 (1993). A presentation of important results from observations with the Spacewatch camera, showing that the terrestrial influx of small (sub–100-meter) asteroids is much higher than previously believed.

Sykes, M.V., and R.G. Walker, "Cometary dust trails. I. Survey," *Icarus*, volume 95, pp. 180–210 (1992). A discussion of the cometary dust trails discovered with the Infra-Red Astronomy Satellite, these trails perhaps having profound implications for terrestrial catastrophism.

Chapter 8

Atkinson, R.J.C., *Stonehenge*, Penguin/Hamish Hamilton, London (1956; revised 1979). The standard archeological account of this famous megalithic structure.

Bauval, R., and A. Gilbert, *The Orion Mystery: Unlocking the Secrets of the Pyramids*, Heinemann, London (1994). The latest word regarding astronomical interpretations of the great pyramids of Egypt.

Castleden, R., *The Making of Stonehenge*, Routledge, London and New York (1993). An excellent up-to-date account of what we know about Stonehenge and its development.

Chippendale, C., *Stonehenge Complete*, Thames and Hudson, London (1983). An excellent, broad discussion of Stonehenge and its context.

Clube, S.V.M., and W.M. Napier, *The Cosmic Winter*, Basil Blackwells, Oxford

(1990). A detailed interpretation of ancient myths, legends, and historical observations in terms of modern astronomical concepts.

Goldsmith, D. (editor), *Scientists Confront Velikovsky*, Cornell University Press, Ithaca and London (1977). All you really need to know about Immanuel Velikovsky's absurd astronomical ideas.

Hawkins, G.S., with J.B. White, *Stonehenge Decoded*, Doubleday and Company, Garden City, New York (1965). The book that started a revolution in archeoastronomy, Hawkins claiming that Stonehenge was built as a gigantic eclipse predictor.

Hawkins, G.S., *Beyond Stonehenge*, Harper and Row, New York (1973). An updated version and expansion of Hawkins' ideas.

Heggie, D.C., *Megalithic Science*, Thames and Hudson, London (1981). An extensive discussion of megalithic monuments, predominantly those of the British Isles.

Hoyle, F., *On Stonehenge*, W.H. Freeman and Company, San Francisco (1977). An accessible discussion of the eclipse and sunrise interpretations of Stonehenge as an astronomical observatory.

Jones, H.D., "Zodiacal light and the pyramids," *Journal of the British Astronomical Association*, volume 100, p. 162 (1990). A brief letter describing the apparent link between the pyramids and the zodiacal light.

Lancaster Brown, P., *Megaliths, Myths and Men: An Introduction to Astro-archaeology*, Blandford Press, Poole, England (1976). An excellent discussion of Stonehenge, other megalithic monuments, and pyramidology.

Chapter 9

Chyba, C., P. Thomas, and K. Zahnle, "The 1908 Tunguska explosion: Atmospheric detonation of a stony asteroid," *Nature*, volume 361, pp. 40–44, (1993). A paper describing a physical model for the entry of large meteoroids into the atmosphere, concentrating on the Tunguska event.

Korobejnikov, V.P., P.I. Chushkin, and L.V. Shurshalov, "Combined simulation of the flight and explosion of a meteoroid in the atmosphere," *Solar System Research*, volume 25, pp. 242–255 (1991). An excellent, detailed model of the interaction between the atmosphere and incoming small asteroids, especially the Tunguska object.

Kresák, L., "The Tunguska object: A fragment of comet Encke?," *Bulletin of the Astronomical Institutes of Czechoslovakia*, volume 29, pp. 129–134 (1978). A discussion of the evidence for the Tunguska projectile's having been part of the Taurid Complex, and derived from Comet Encke.

Welfare, S., and J. Fairley, "The great Siberian explosion," pp. 153–167, in *Arthur C. Clarke's Mysterious World*, Collins, London (1980). A popular account of the Tunguska explosion, concluding with Clarke's suggestion that the Tunguska object might have been a large body in the Beta Taurid meteor shower.

Zahnle, K., "Airburst origin of dark shadows on Venus," *Journal of Geophysical Research*, volume 97, pp. 10243–10255 (1992). A discussion of the cratering record of Venus as implied by the *Magellan* radar images, and why there are few craters smaller than about 8 kilometers in diameter on that planet, due to its thick atmosphere.

Chapter 10

Gehrels, T., "Scanning with charge-coupled devices," *Space Science Reviews*, volume 58, pp. 347–375 (1991). A detailed discussion of how the Spacewatch scanning system works.

Helin, E.F., and E.M. Shoemaker, "Palomar Planet-Crossing Asteroid Survey 1973–1978," *Icarus*, volume 40, pp. 321–328 (1979). The early history of this important program.

Helin, E.F., and R.S. Dunbar, "Search techniques for near-Earth asteroids," *Vistas in Astronomy*, volume 33, pp. 21–37 (1990). A compact review of near-Earth asteroid searches through to 1989.

Shoemaker, C.S., and E.M. Shoemaker, "The Palomar Asteroid and Comet Survey (PACS) 1982–1987," *Lunar and Planetary Science*, volume XIX, pp. 1077–1078 (1988). Results from the early years of this program.

Steel, D., and R.H. McNaught, "The Anglo-Australian Near-Earth Asteroid Survey," *Australian Journal of Astronomy*, volume 4, pp. 42–46 (1991). A description of the foundation of the Australian program.

Chapter 11

Clarke, Arthur C., *The Hammer of God*, Bantam Books, New York (1993). A fictional account of the implementation (and utility) of the Spaceguard program, the name of which was inspired by Clarke's earlier science fiction novel, *Rendezvous with Rama* (1973).

Morrison, D. (editor), *Report of the NASA Near-Earth-Object Detection Workshop: The Spaceguard Survey*, NASA/Jet Propulsion Laboratory, Pasadena, California (1992).

Chapter 12

Ahrens, T.J., and A.W. Harris, "Deflection and fragmentation of near-Earth asteroids," *Nature*, volume 360, pp. 429–433 (1992). A technical description of how to deflect asteroids perceived as an impact hazard by using stand-off nuclear explosives.

Canavan, G.H., J.C. Solem, and J.D.G. Rather, *Proceedings of the Near-Earth-Object Interception Workshop*, Los Alamos National Laboratory, Los Alamos, New Mexico, publication LA-12476-C (1993). A collection of papers presented at the meeting of the Interception Committee at Los Alamos in January 1992, along with summaries of the subcommittees set up to report on various specific facets of the problem of interception and deflection.

Harris, A.W., G.H. Canavan, C. Sagan, and S.J. Ostro, "The deflection dilemma: Use versus misuse of technologies for avoiding interplanetary collision hazards," pp. 1145–1156, in *Hazards due to Comets and Asteroids*, edited by T. Gehrels, University of Arizona Press, Tucson (1994). A discussion of the ways in which the ability to defend the Earth against impacts also allows the use of that capability for offensive purposes (for example, to deflect an asteroid *toward* an enemy nation).

Kleiman, L.A. (editor), *Project Icarus*, MIT Report Number 13, MIT Press, Cambridge, Massachusetts (1968). The report of the MIT students who were set the task of diverting asteroid 1566 Icarus, which was imagined to be on a collision course with our planet.

Marsden, B.G., and D.I. Steel, "Warning times and impact probabilities for long-period comets," pp. 221–239, in *Hazards due to Comets and Asteroids*, edited by T. Gehrels, University of Arizona Press, Tucson (1994). A discussion of the possible threat from long-period/parabolic comets that cannot be discovered prior to the orbit on which they might impact the Earth. If Spaceguard went ahead, these comets would be the only large (bigger than 1 kilometer) objects for which we would have limited advance warning.

Melosh, H.J., and I.V. Nemchinov, "Solar asteroid diversion," *Nature*, volume 366, pp. 21–22 (1993). A description of a technique for deflecting asteroids by the use of solar sails. The feasibility of this idea has been questioned by some scientists, but it does provide a possible alternative to the much-hated nuclear weapon deflection method.

Rather, J.D.G., J.H. Rahe, and G.H. Canavan, *Summary Report of the Near-Earth-Object Interception Workshop*, NASA Headquarters, Washington, D.C. (1992). A summary of the major conclusions from the various subcommittees of the Interception Workshop held at Los Alamos, January 1992.

Sagan, C., and S.J. Ostro, "Dangers of asteroid deflection," *Nature*, volume 368, p. 501 (1994). A brief summary of the concerns detailed in the book chapter by Harris et al. in this section.

Index

Note: Comets are listed by their specific names, i.e., Comet Halley appears as Halley, Comet. Page numbers followed by *f* indicate figures; page number followed by *n* indicate notes; page numbers followed by *t* indicate tables.